E-MRS 2011 Fall Symposium I: Advances in Transparent Electronics, from Materials to Devices III

AA002602

IOP Conference Series: Materials Science and Engineering Volume 34

Warsaw, Poland
19 – 23 September 2011

ISBN: 978-1-63439-845-9
ISSN: 1757-8981

E-MRS 2011 Fall Symposium I: Advances in Transparent Electronics, from Materials to Devices III

IOP Conference Series: Materials Science and Engineering Volume 34

Warsaw, Poland
19 – 23 September 2011

ISBN: 978-1-63439-845-9
ISSN: 1757-8981

IOP Institute of Physics

Printed from e-media with permission by:

Curran Associates, Inc.
57 Morehouse Lane
Red Hook, NY 12571

Some format issues inherent in the e-media version may also appear in this print version.

Copyright© by the Institute of Physics
All rights reserved.

Printed by Curran Associates, Inc. (2015)

For permission requests, please contact the Institute of Physics
at the address below.

Institute of Physics
Dirac House, Temple Back
Bristol BS1 6BE UK

Phone: 44 1 17 929 7481
Fax: 44 1 17 920 0979

techtracking@iop.org

Additional copies of this publication are available from:

Curran Associates, Inc.
57 Morehouse Lane
Red Hook, NY 12571 USA
Phone: 845-758-0400
Fax: 845-758-2634
Email: curran@proceedings.com
Web: www.proceedings.com

TABLE OF CONTENTS

IOP Conf. Series: Materials Science and Engineering **34** (2012) 011001 doi:10.1088/1757-899X/34/1/011001

Preface

Symposium I / E-MRS Fall Meeting
September 2011, Warsaw

As world-wide interest in Transparent Electronics appears to be exploiting, there is a need for intense information exchange and platforms for debate in order to facilitate rapid advances on both the materials side and in device fabrication.

Symposium I complemented a highly successful series of symposia in this field, organized over the past few years by E-MRS, with the aim of contributing to the present rush of new knowledge and advances in the so-called '*post-Si electronics era*'. This Symposium was organized by a team of international leading experts, ranging from materials scientists to electronic engineers with a solid background in both the scientific and technological fields addressed.

The scope of this symposium, which was to attract young scientists and engineers interested in the field, who would have a first class opportunity to interact with a pool of prominent invited speakers, in subjects ranging from materials science to device fabrication and applications, was fully accomplished.

Published under licence by IOP Publishing Ltd

Impurity-doped ZnO Thin Films Prepared by Physical Deposition Methods Appropriate for Transparent Electrode Applications in Thin-film Solar Cells

Tadatsugu Minami, Toshihiro Miyata and Jun-ichi Nomoto

Optoelectronic Device System R&D Center, Kanazawa Institute of Technology, 7-1 Ohgigaoka, Nonoichi, Ishikawa 921-8501, Japan

minami@neptune.kanazawa-it.ac.jp

Abstract. This paper describes the development of transparent conducting impurity-doped ZnO thin films that would be appropriate for applications as transparent electrodes in thin-film solar cells. Transparent conducting Al-, B- and Ga-doped ZnO (AZO, BZO and GZO) thin films were prepared in a thickness range from 500 to 2000 nm on glass substrates at 200°C using various physical deposition methods: BZO films with vacuum arc plasma evaporation, AZO and GZO films with different types of magnetron sputtering depositions (MSDs) and all films with pulsed laser deposition (PLD). The suitability and stability of the electrical properties and, in addition, the suitability of the light scattering characteristics and surface texture formation were investigated in the prepared thin films. In particular, the suitability and stability evaluation was focused on the use of AZO, BZO and GZO thin films prepared by doping each impurity at an appropriate content to attain the lowest resistivity. The higher Hall mobility obtained in impurity-doped ZnO thin films with a resistivity on the order of 10^{-4} Ωcm was related more to the content, *i.e.*, the obtained carrier concentration, rather than the kind of impurity doped into the films. The stability of resistivity of the BZO thin films in long-term moisture-resistance tests (in air at 85% relative humidity and 85°C) was found to be lower than that of the AZO and GZO thin films. The surface texture formation was carried out by wet-chemical etching (in a 0.1% HCl solution at 25°C) conducted either before or after being heat-treated either with rapid thermal annealing (RTA) or without RTA. The suitability of the light scattering characteristics and the surface texture formation obtainable by wet-chemical etching (for use in transparent electrode applications) was considerably dependent on the deposition method used as well as whether the wet-chemical etching was conducted with or without RTA. A significant improvement of both transmittance and haze value at wavelengths up to about 1200 nm in the near-infrared region was attained in surface-textured AZO films that were prepared by r.f. power superimposed d.c. MSD as well as etched after being heat treated with RTA at 500°C for 5 min in air. The obtained suitability and stability in impurity-doped ZnO thin films were related more to the content rather than the kind of impurity doped into the films as well as to the deposition method used.

1. Introduction

Recently, highly transparent and conductive impurity-doped ZnO thin films such as Al- and Ga-doped ZnO (AZO and GZO) prepared by magnetron sputtering depositions (MSDs) have been attracting much attention as alternatives to the indium-tin-oxide transparent electrodes used in flat panel displays

Published under licence by IOP Publishing Ltd

(FPDs) such as liquid crystal and organic electroluminescent displays [1-4]. For such applications, it is necessary that lower resistivity AZO and GZO thin films be prepared with a thickness below approximately 200 nm on substrates at a temperature below approximately 200°C. The lower resistivity can be achieved in AZO and GZO thin films by doping the impurities with an Al content (Al/(Al+Zn) atomic ratio) of 2-4 at.% and a Ga content (Ga/(Ga+Zn) atomic ratio) of 5-8 at.%, irrespective of the deposition method used [1, 2, 5-8]. At the present time, AZO and B-doped ZnO (BZO) thin films are in practical use for transparent electrode applications in $CuIn_{1-X}Ga_XSe_2$-based thin-film solar cells [9-12]. In addition, impurity-doped ZnO thin films, such as AZO, GZO and BZO with a textured surface structure as well as a high transmittance in the near-infrared region, have recently attracted much attention for transparent electrode applications in Si-based thin-film solar cells [13-24]. It is necessary to form impurity-doped ZnO thin films with a doubly textured surface structure that can effectively scatter the incident visible and near-infrared light [25]. Also, there may be a significant reduction of transmittance in the near-infrared region that is caused mainly by reflectance as well as absorption attributed to free electrons in the highly transparent and conductive thin films resonantly interacting with electromagnetic waves (electron plasma resonance). Thus, impurity-doped ZnO thin films that would be suitable for transparent electrode applications in thin-film solar cells must necessarily attain not only a decrease of plasma resonance frequency by lowering the carrier concentration while retaining a resistivity on the order of 10^{-4} Ωcm, but also a significant scattering of light incident on the film by surface texturing. This suggests that the lower resistivity AZO and GZO thin films that were developed for transparent electrode applications in FPDs, as described above, may not be appropriate for transparent electrode applications in thin-film solar cells. In addition, it is known that the electrical and structural properties obtainable in polycrystalline impurity-doped ZnO thin films are considerably affected by the deposition conditions and method used [1, 7, 8, 26, 27].

This paper describes the development of transparent conducting impurity-doped ZnO thin films that would be appropriate for applications as transparent electrodes in thin-film solar cells. The suitability and stability of the electrical properties as well as the suitability of the light scattering characteristics and the surface texture formation obtainable by wet-chemical etching are evaluated for transparent electrode applications. Transparent conducting AZO, BZO and GZO thin films were prepared with a thickness above 500 nm on glass substrates at 200°C using pulsed laser deposition (PLD), vacuum arc plasma evaporation (VAPE) and different types of magnetron sputtering depositions (MSDs). The evaluation of suitability and stability is focused on the use of AZO, BZO and GZO thin films prepared by doping each impurity at an appropriate content, *i.e.*, resulting in the lowest resistivity as well as most suitable for transparent electrode applications in FPDs attainable.

2. Experimental

The impurity-doped ZnO thin films were prepared on glass (OA-10, Nippon Electric Glass Co., Ltd.) at a substrate temperature of 200°C using three kinds of MSD as well as PLD and VAPE methods. AZO and GZO thin films were prepared with three types of MSD using a magnetron sputtering apparatus (SIH-300RD, ULVAC, Inc.) with a sintered oxide target; the three types used a d.c. and an r.f. (13.56 MHz) power supply applied either separately or in combination [28]. The oxide targets used were commercially available high-density-sintered rectangular AZO and GZO targets (127 mm×275 mm) prepared with an Al_2O_3 content of 2 wt.% and a Ga_2O_3 content of 5.7 wt.%, respectively. The d.c. and r.f. MSDs (dc- and rf-MSDs) were conducted with a d.c. power of 700 W and an r.f. power of 1000 W, respectively. The dc-MSDs that incorporated r.f. power (rf+dc-MSD) were carried out by adding a 700 W r.f. component to a constant dc power of 570 W. In both the prepared AZO and GZO films, the resulting deposition rates were roughly 100, 20 and 100 nm/min with dc-MSD, rf-MSD and rf+dc-MSD, respectively. The substrate, area of 200 mm×200 mm, was placed parallel to the target surface at a minimum substrate-target distance of 90 mm, as shown in Fig. 1. Sputtering depositions were carried out in a pure Ar gas atmosphere at a pressure of 0.2 Pa. The substrate was moving at a velocity of 200 mm/min during the depositions. In film depositions on moving substrates, the resulting film thickness is determined by the number of times the substrate, with its reciprocating motion,

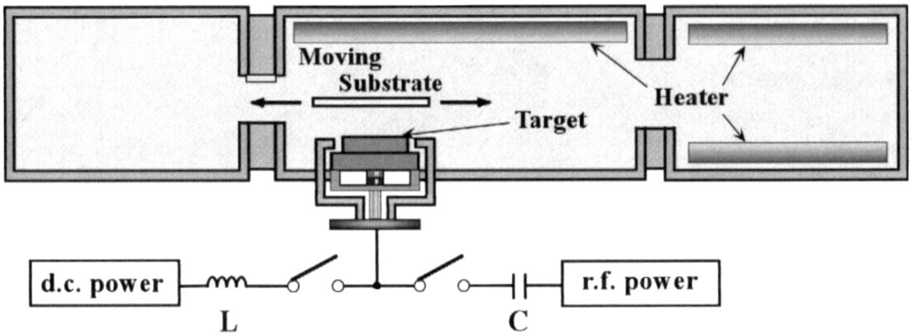

Fig. 1 Schematic diagram of MSD apparatus and arrangement of the target and substrate.

passes above the target surface, as shown in Fig. 1. With PLD, the AZO, GZO and BZO thin films were prepared using an ArF excimer laser (wavelength, 193 nm; repetition rate, 20 Hz; pulse width, 20 ns; and fluence, 350 mJ/cm^2) under the following deposition conditions: pressure, vacuum below 4.0×10^{-4} Pa; target-substrate distance, 40 mm; and target, sintered AZO, GZO or BZO pellet [29]. The AZO and GZO pellets, used in PLD, were prepared by cutting the commercially available sintered AZO and GZO targets used in the MSDs. With VAPE, the BZO thin films were prepared at a pressure of 0.15 Pa, an Ar gas flow rate of 20 sccm and a cathode plasma power of 4.5 kW by a VAPE apparatus using pellets and fragments of sintered BZO as the source material [8, 30]. The BZO pellets and fragments, used in both PLD and VAPE, were prepared by cold pressing a mixture of powdered ZnO and B_2O_3 dopant (with contents of 0.1-1.3 wt.%), followed by sintering in an Ar gas atmosphere at 1000°C.

In this work, the thickness of all deposited thin films was in the range from 500 to 2000 nm, irrespective of the deposition method used. In moisture-resistance tests, changes in the electrical properties of impurity-doped ZnO thin films were measured over time with exposure to highly moist environments: air at 85% relative humidity and 85°C [8, 31]. The suitability of the light scattering characteristics for thin-film solar cell applications was evaluated by carrying out surface texturing of the samples with wet-chemical etching in a 0.1% HCl solution at 25°C conducted either before or after heat treatment with rapid thermal annealing (RTA) as well as without RTA treatment [32, 33]. The RTA treatment was carried out using a lamp heated furnace at a temperature of 400 or 500°C for 3 or 5 min in air. Subsequently, observations of surface morphology and measurements of the optical transmittance and the diffusive component of the surface-textured impurity-doped ZnO thin films were also performed. The Al, Ga and B content doped into the deposited thin films was evaluated by x-ray photoelectron spectroscopy. Film thickness was measured using a conventional surface roughness detector with stylus (Tokyo Seimitsu Surfcom MD-S69A). The following evaluations were conducted on the deposited thin films: analyses of the crystallographic properties by X-ray diffraction (XRD) using a conventional X-ray unit (Rigaku Ultima-2100) with a copper anode and measurements of the optical and electrical properties using a spectrophotometer (Hitachi U-3500) and the van der Pauw method (Accent HL-5500PC), respectively.

3. Results and Discussion

3.1. Influence of doped impurity and deposition method on electrical properties

Transparent conducting BZO and AZO thin films prepared with a high deposition rate on substrates at temperatures below approximately 200°C (in practical use for transparent electrode applications in $CuIn_{1-X}Ga_XSe_2$-based thin-film solar cells, mentioned above) have been prepared by a low pressure metalorganic chemical deposition and rf-MSD methods, respectively [9-12]. However, the various properties obtainable in the polycrystalline impurity-doped ZnO thin films prepared on substrates at a low temperature were affected by the film thickness. In addition, various properties not only improved

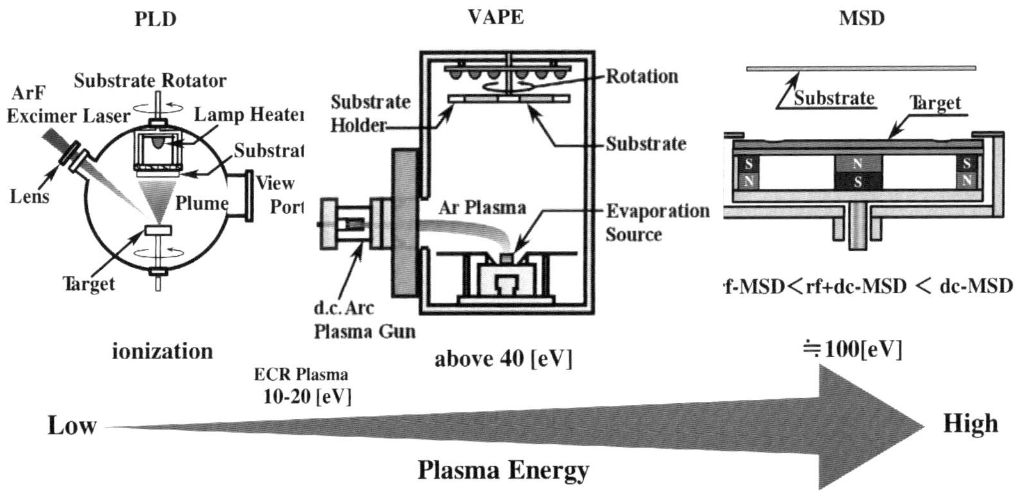

Fig. 2 Comparison of various deposition methods and plasma energy during the deposition.

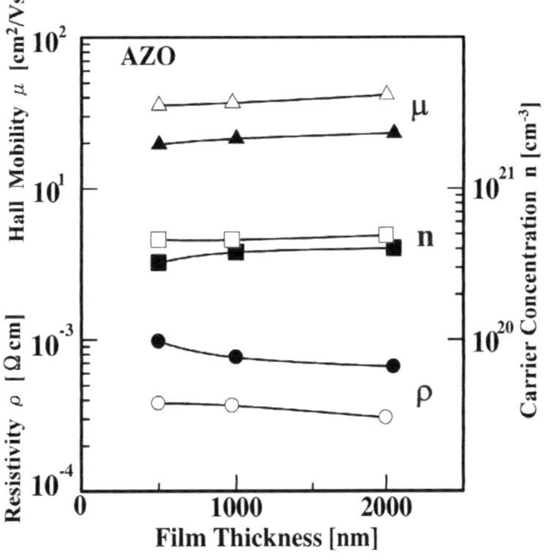

Fig.3 Electrical properties as functions of thickness for AZO thin films prepared by dc-MSD (solid) and rf+dc-MSD (open).

significantly as the thickness was increased up to approximately 500 nm, but also were considerably affected by the kind and content of doped impurity as well as the kind of deposition method used. In this work, therefore, transparent conducting impurity-doped ZnO thin films that would be the most suitable for applications as transparent electrodes in thin-film solar cells were investigated in AZO, BZO and GZO thin films prepared with a thickness above 500 nm using various physical deposition methods, *i.e.*, PLD, VAPE and MSDs utilizing plasma generated with different energies, as shown in Fig. 2. As an example, the electrical properties as functions of thickness are shown in Fig. 3 for transparent conducting AZO thin films prepared with a thickness above 500 nm on substrates at 200°C by different types of MSD using an MSD apparatus with a sintered AZO target. The obtained resistivity exhibited a tendency to decrease slightly as the thickness was increased from 500 to about 2000 nm, irrespective of the MSD method used. However, it should be noted that the obtainable resistivity in AZO thin films was considerably dependent on the deposition method used; the obtainable resistivity in AZO thin films prepared by dc-MSD was always higher than that in films prepared with the same thickness by rf+dc-MSD. It was also found that the obtained resistivity and its thickness dependence in AZO thin films prepared by rf-MSD were almost equal to those in thin films prepared by rf+dc-MSD. However, the obtained deposition rate in thin films prepared by rf-MSD was very much lower than that found in dc-MSD or rf+dc-MSD, as mentioned above, possibly too low to use practically for transparent electrode applications in thin-film solar cells [28, 33].

The obtained lower resistivity in the AZO thin films prepared by rf+dc-MSD and rf-MSD, relative to dc-MSD, is related mainly to a higher mobility that may be attributable to better crystallinity of the films prepared by rf+dc-MSD and rf-MSD, as evidenced by XRD analyses [32]. In addition, it is well known that AZO thin films prepared on low temperature substrates by MSDs exhibit a resistivity distribution on the substrate surface that corresponds to the erosion pattern on the target surface, if the thin films were prepared with the substrate position fixed relative to that of the target [28, 32]. With this configuration, the resistivity increase at the location on the substrate surface that corresponds to the erosion area of the target in films prepared by dc-MSD was always higher than that in films prepared by rf-MSD or rf+dc-MSD [32]. The increase in resistivity is attributable to an enhancement in the amount and activity of oxygen that reaches the location on the substrate surface corresponding to the erosion area of the target, resulting from negatively ionized oxygen accelerated by the higher d.c. sputter voltage in dc-MSD than in rf-MSD or rf+dc-MSD [28, 32, 34-39]. In the current work using moving substrates during the deposition, the amount of resistivity obtainable in the thin films exhibits an average value that is related to the resistivity distribution that results from depositions on the surface of fixed substrates. Thus, the higher resistivity obtained in the AZO thin films prepared by dc-MSD shown in Fig. 3 is mainly attributed to the increase in the level of the resulting resistivity distribution on the substrate surface described above. However, it should be noted that the resulting resistivity distribution on the substrate as well as the obtainable thickness dependence of resistivity in the thin films prepared by the MSDs were considerably affected by differences that can be associated with the target used, *e.g.*, the kind of doped impurity and supplier of the target [7, 8, 40-42]. In GZO thin films prepared with a thickness above 500 nm on substrates at 200°C by dc-MSD and rf+dc-MSD, the influence of the deposition method used on the obtained electrical properties in GZO films differed from that found in AZO films described above. However, in regard to thickness dependence, the obtained resistivity in the prepared GZO films was found to exhibit a tendency to decrease slightly as the thickness was increased from 500 to about 2000 nm, which is similar to that found in AZO films, irrespective of the MSD method used. In GZO and AZO films prepared by different MSDs, any differences found concerning the influence of the deposition method used on the obtained electrical properties were mainly related to differences in the impurity content doped into the films, *i.e.*, the obtained carrier concentration. Table 1 summarizes the obtained electrical properties and the doped impurity content in AZO and GZO thin films prepared with high deposition rates and a thickness of 500 or 2000 nm on glass substrates at 200°C by dc-MSD or rf+dc-MSD using an AZO or GZO target with an Al_2O_3 content of 2 wt.% or a Ga_2O_3 content of 5.7 wt.%, respectively, resulting in the lowest resistivity as well as the most suitable for transparent electrode applications in FPDs attainable [1, 2, 5-8].

Table 1 Electrical properties and impurity content obtained in AZO and GZO thin films prepared by dc-MSD or rf+dc-MSD.

Target	Deposition Method	Impurity Content[at.%]	Thickness: 500nm			Thickness: 2 μm		
			ρ [Ωcm]	μ [cm²/Vs]	n[cm⁻³]	ρ [Ωcm]	μ [cm²/Vs]	n[cm⁻³]
AZO	dc-MSD	4.6	9.73×10^{-4}	19.8	3.24×10^{20}	5.20×10^{-4}	25.6	4.70×10^{20}
	rf+dc-MSD		3.69×10^{-4}	29.7	5.70×10^{20}	3.06×10^{-4}	41.7	4.89×10^{20}
GZO	dc-MSD	7.3	3.55×10^{-4}	24.1	7.31×10^{20}	3.00×10^{-4}	28.7	7.27×10^{20}
	rf+dc-MSD		3.77×10^{-4}	31.1	5.32×10^{20}	3.38×10^{-4}	35.3	5.23×10^{20}

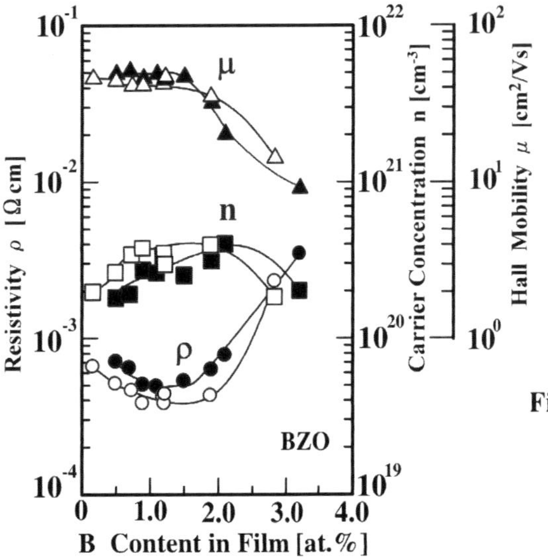

Fig.4 Electrical properties as functions of B-doping content for BZO thin films prepared with a thickness of approximately 500 nm by PLD (open) and VAPE (solid).

Table 2 Electrical properties and impurity content obtained in impurity-doped ZnO thin films prepared by PLD.

	BZO	AZO	GZO
Impurity Content [at.%]	1.23	2.46	8.07
ρ [Ω cm]	4.45×10^{-4}	2.51×10^{-4}	2.81×10^{-4}
μ [cm²/Vs]	47.2	32.5	22.5
n [cm^{-3}]	2.97×10^{20}	7.64×10^{20}	9.85×10^{20}

In contrast, an optimal B content doped into transparent conducting BZO thin films that can obtain the lowest resistivity may be not satisfactorily decided yet. The resistivity, Hall mobility (μ) and carrier concentration (n) as functions of B content are shown in Fig. 4 for BZO thin films prepared with a thickness of 500 nm on substrates at 200°C using different deposition methods [8]. The Hall mobility maintained a high value above 40 cm^2/Vs as the B content was increased up to approximately 1.5 at.%, and then it decreased markedly with any further increase of the B content; the carrier concentration increased as the B content was increased up to approximately 2 at.%, irrespective of the deposition method used. As can be seen in Fig. 4, the obtained electrical properties in BZO films, even when prepared with the same B doping content, were affected by the deposition method used; the obtainable resistivity in BZO films prepared by PLD (open data) is always lower than that in films prepared by VAPE (solid data). It has been also reported that the obtainable resistivity in transparent conducting AZO and GZO thin films prepared on glass substrates at a low temperature by PLD was lower than that in films prepared by other deposition methods [1, 7, 26]. Similarly, this was also confirmed in this work, namely, that the obtainable resistivity in transparent conducting AZO and GZO thin films prepared on glass substrates at 200°C by PLD was lower than that in films prepared by MSDs, as shown in Table 1. Table 2 summarizes the obtained electrical properties and doped impurity content in AZO, BZO and GZO thin films prepared with a thickness of 500 nm on substrates at 200 °C by PLD. The Hall mobility of 47.2 cm^2/Vs obtained in the BZO film was higher than that in the AZO

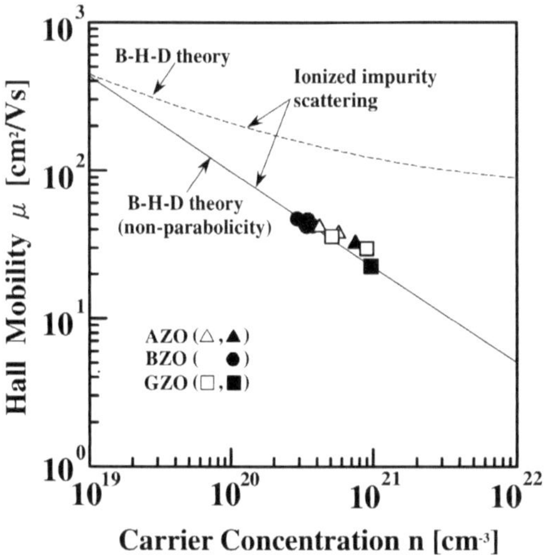

Fig.5 The μ -n relationship obtained in AZO (△, ▲), BZO (●) and GZO (□,■) films prepared with a thickness above 500 nm by PLD (solid) and rf+dc-MSD (open). The solid line represents the ionized impurity scattering μ -n relationship calculated using B-H-D theory with nonparabolicity.

and GZO films. However, the obtained resistivity in the BZO film was higher than that in the AZO and GZO films, because the obtained carrier concentration in the BZO film was lower.

Figure 5 shows the relationship between μ and n in high-mobility AZO (triangle), BZO (circle) and GZO (square) thin films prepared on glass substrates at 200°C by different deposition methods. The BZO films were prepared with a thickness of 500 nm by PLD (solid), and both the AZO and GZO thin films were prepared with a thickness of 500 nm by PLD (solid) and 2000 nm by rf+dc-MSD (open), as shown in Tables 2 and 1, respectively. Two theoretical μ-n curves calculated using the ionized impurity scattering theory reported by Brooks-Herring-Dingle (B-H-D) (dashed line) and the B-H-D theory with nonparabolicity of the conduction band taken into account (solid line) [43-45] are also indicated in Fig. 5. As can be seen in Fig. 5, all the plotted data points lie on the theoretical curve calculated using the latter B-H-D theory. This suggests that the obtained high mobility in BZO films is attributable to the lower carrier concentration, because the electrical conduction mechanism in all the AZO, BZO and GZO thin films is mainly dominated by the ionized impurity scattering. In addition, the obtained lower carrier concentration in BZO films is attributable to a lower content of B doped into the BZO film than the content of Al doped into the AZO or Ga doped into the GZO films, as seen in Table 2. It should be noted that in AZO and GZO films the resistivity obtained in films prepared with a thickness of 500 nm by PLD was approximately the same as that obtained in films prepared with a thickness above approximately 2000 nm by rf+dc-MSD. This difference between the different deposition methods may be attributable mainly to the crystallinity of the prepared films, resulting from the difference in the amount and activity of oxygen reaching the substrate surface that is also related to differences in plasma energy during the deposition, as shown in Fig. 2. Figure 6 shows typical XRD patterns of AZO thin films prepared with a thickness of 1000 nm on glass substrates at 200°C by various deposition methods: PLD, dc-MSD and rf+dc-MSD. All the prepared AZO films had their c-axis strongly oriented perpendicular to the substrate surface; however, the crystallinity estimated from the intensity and half width of the (0002) XRD peak was considerably dependent on the deposition method used. As can be seen in Fig. 6, the extent of orientation preference as well as the position of the observed (0002) XRD peak were strongly dependent on the deposition method used. The (0002) XRD peak in highly conductive and transparent AZO thin films prepared on glass substrates at temperatures below approximately 200°C was always observed at higher angles than that in standard ZnO powder, irrespective of the deposition method used. This is consistent with an XRD peak position shift that not only was uncorrelated to the c-axis orientation, but also was attributable mainly to the difference of ion radius between Zn^{2+} and Al^{3+}, as published previously in many reports [46-48]. However, it should be noted that AZO thin films prepared with a higher c-axis orientation exhibited a

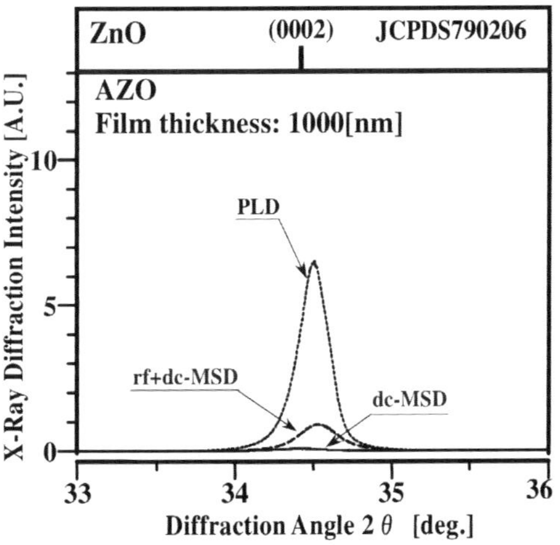

Fig.6 XRD patterns for as-deposited AZO films prepared with a thickness of 1000 nm by PLD (dotted line), dc-MSD (solid line) and rf+dc-MSD (dashed line).

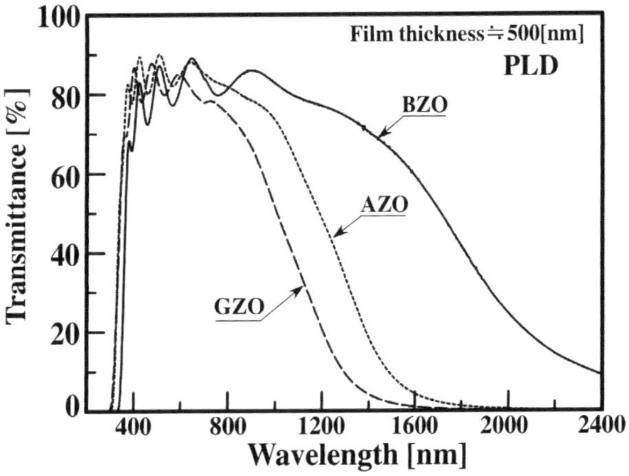

Fig.7 Typical transmission spectra for AZO (dotted line), BZO (solid line) and GZO (dashed line) films prepared with a thickness of 500 nm on glass substrates by PLD.

tendency to have a higher Hall mobility. It was also found that the orientation of AZO thin films prepared by MSDs improved considerably with increasing thickness, even at thicknesses above 500 nm, whereas that of films prepared by PLD almost saturated as the thickness was increased above 500 nm. Thus, the difference between the resulting thickness dependence of Hall mobility (or resistivity) in AZO films prepared by rf+dc-MSD and that in AZO films prepared by PLD, as seen in Figs. 3-5 and Tables 1 and 2, may be attributable mainly to differences in the crystallinity, such as c-axis orientation, found among the different deposition methods, as mentioned above. However, it should be noted that the obtainable Hall mobility in GZO films was mainly dependent on the difference in the crystallinity as well as the higher content of Ga doped into the films relative to the content of the impurity doped into the AZO and BZO films.

Typical transmission spectra are shown in Fig. 7 for the AZO, BZO and GZO thin films prepared with a thickness of 500 nm on OA-10 glass substrates shown in Table 2. High averaged transmittance above approximately 80% in the visible range was obtained in all the thin films; however, there were differences observed among the spectra exhibited by the thin films. The red-shift of the absorption edge and the extended transmission to a longer wavelength in the near-infrared region observed from the BZO thin film relative to data from the AZO and GZO thin films are attributable to the Burstein-Moss effect and the decrease of resonance plasma frequency, respectively, resulting from the lower carrier concentration obtained in the BZO film. The decrease in carrier concentration, BZO, AZO and GZO in descending order, is mainly related to the impurity content doped into the films, as seen in Table 2.

3. 2. Light scattering characteristics obtainable by surface texturing

3.2.1 Influence of deposition method and RTA treatment

The suitability of the light scattering characteristics for thin-film solar cell applications was evaluated by carrying out surface texturing of impurity-doped ZnO thin films with wet-chemical etching conducted either before or after heat treatment with RTA as well as without RTA treatment. To evaluate the light scattering characteristics of impurity-doped ZnO thin films with a surface textured by wet-chemically etching, the optical transmittance and the diffusive component of the films were measured; the haze value was estimated from these measured transmittances [11, 18, 20, 22, 32, 33]. It

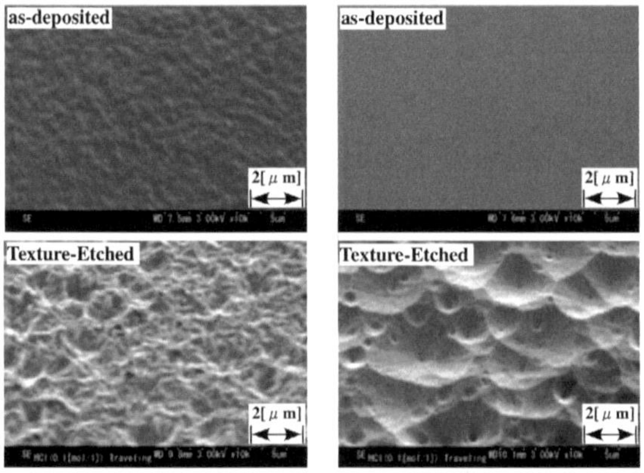

(a) dc-MSD **(b) rf+dc-MSD**

Fig.8 SEM images of the surface of as-deposited and texture-etched AZO films prepared by (a) dc-MSD and (b) rf+dc-MSD.

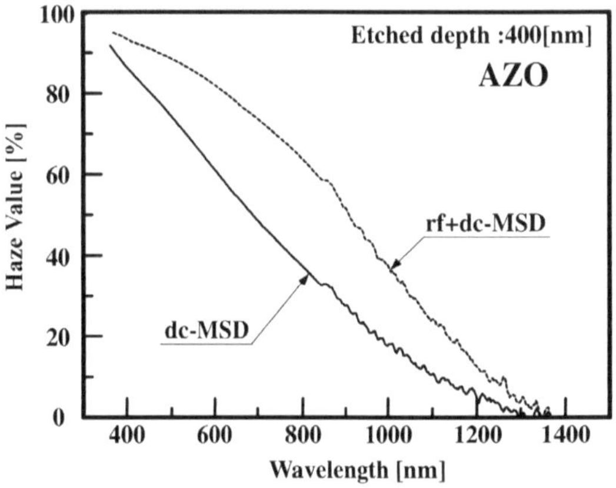

Fig.9 Haze spectra of texture-etched AZO films prepared by dc-MSD and rf+dc-MSD.

was found that the obtainable light scattering characteristics in impurity-doped ZnO thin films with a textured surface structure were considerably affected by the thickness and the doped impurity as well as the kind of deposition method used. As an example of the relationship between the resulting texture structure and the deposition method used, SEM images observed before (as-deposited) and after chemical etching are shown in Figs. 8(a) and 8(b) for AZO thin films prepared by dc-MSD and rf+dc-MSD methods, respectively. AZO films prepared with a thickness of 2000 nm, as shown in Table 1, were used for surface texturing, because a thickness above about 500 nm is appropriate for AZO films in practical use for transparent electrode applications in thin-film solar cells. The surface-textured AZO films were etched to a depth of approximately 400 nm with an etching rate that was dependent on the type of MSD used; typical etching rates of AZO films prepared by dc-MSD and rf+dc-MSD were approximately 11.4 and 4.3 nm/min, respectively. The surface image observed from the texture-etched AZO film prepared by rf+dc-MSD exhibited larger etch pits than those observed from the film prepared by dc-MSD, whereas the surface of the as-deposited film prepared by rf+dc-MSD was always smoother than that prepared by dc-MSD. Figure 9 shows the haze spectra of the surface-textured AZO films prepared by dc-MSD and rf+dc-MSD and etched to a depth of approximately 400 nm shown in Fig. 8. The obtained haze value in the film prepared by rf+dc-MSD is considerably higher in the range from long wavelengths in the visible region to the near-infrared region than that obtained in the film prepared by dc-MSD. The difference of the haze spectrum and haze value obtainable between texture-etched AZO thin films prepared by dc-MSD and rf+dc-MSD may be mainly attributable to the difference in the texture surface structure shown in Fig. 8. A broad haze spectrum and a haze value as low as 40% at a wavelength of 1000 nm were obtained in the texture-etched AZO thin film prepared by rf+dc-MSD. In addition, the influence of the type of deposition method used on the obtainable light scattering characteristics in surface-textured GZO thin films prepared by dc-MSD and rf+dc-MSD and wet-chemically etched was found to be the same as that found in AZO films, as described above (detailed data not shown here). However, the obtained haze spectra, even in texture-etched AZO and GZO thin films prepared by rf+dc-MSD, may not be appropriate for transparent electrode applications in Si-based thin-film solar cells fabricated with a micro-crystalline $Si_{1-X}Ge_X$ thin-film active layer that exhibits its maximum photo response at a wavelength near 1000 nm [49].

We have reported that the obtainable haze value in AZO thin films was considerably improved by wet-chemically etching the films after heat treatment with RTA at a temperature above approximately 400°C in air [33]. However, the order that the RTA treatment and the etching were

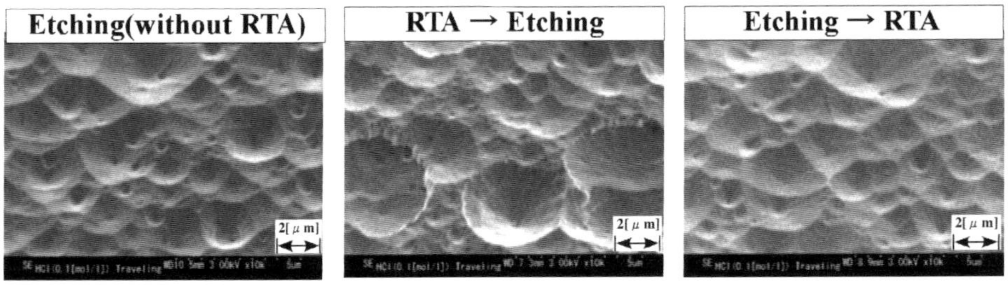

(a) without RTA **(b) after RTA** **(c) before RTA**

Fig.10 SEM images of the surface of surface-textured AZO films prepared by rf+dc-MSD and etched (a) without RTA and (b) after and (c) before being heat treated with RTA.

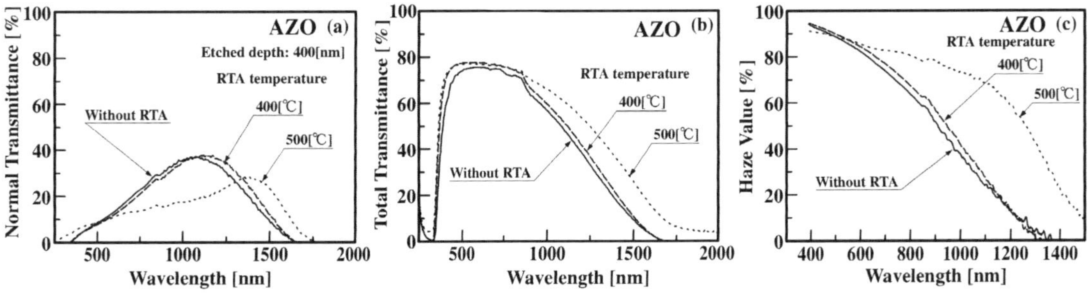

Fig.11 (a) Normal, (b) total transmission and (c) haze spectra for surface-textured AZO films prepared by rf+dc-MSD and etched without RTA (solid lines) and etched after being heat treated with RTA at a temperature of 400 (dashed lines) and 500°C (dotted lines) for 3 min.

conducted considerably affected the obtainable surface texture [33]. Conducting the etching after a heat treatment with RTA in air resulted in larger size etch pits as well as higher haze values than were obtained in AZO films that were etched before the RTA, irrespective of the type of MSD used. Examples of SEM images of the surface of surface-textured AZO films that were (a) etched without RTA treatment as well as either (b) after or (c) before RTA treatment at a temperature of 500°C for 3 min in air are shown in Fig. 10. The etching was carried out to a depth of approximately 400 nm using AZO films prepared with a thickness of 2000 nm by rf+dc-MSD, as shown in Table 1. The etched depth of approximately 400 nm was found to form a suitable textured surface when using AZO thin films with a thickness of 2000 nm; the obtained size of etch pits was insufficient with an etched depth less than 400 nm, whereas an etched depth over 400 nm reached the glass surface. It was also found that AZO thin films prepared by rf+dc-MSD and etched after heat treatment with RTA always exhibited larger etch pits than those observed in equivalent films prepared by dc-MSD, *i.e.*, exhibited enhanced surface textures. The variations seen in the etch pit characteristics (texture sizes) described above, even with the etch carried out under the same wet-chemical etching conditions, may be attributable mainly to differences in the crystallinity, such as c-axis orientation, resulting from the type of MSD used and whether an RTA treatment was applied as well as the RTA conditions [33]. Typical normal and total transmission spectra and haze spectra are shown in Figs. 11(a), (b) and (c), respectively, for various surface-textured AZO films (to a depth of 400 nm) prepared with a thickness of 2000 nm by rf+dc-MSD and etched without RTA treatment and etched after being heat treated with RTA at a temperature of 400 and 500°C for 3 min. It should be noted that AZO thin films heat treated with RTA at a temperature of 500°C for 3 min exhibited an increase of transmittance in the near-infrared region at wavelengths longer than 1000 nm. It was also found that AZO thin films prepared by rf+dc-MSD always exhibited larger haze values than those obtained in equivalent films prepared by

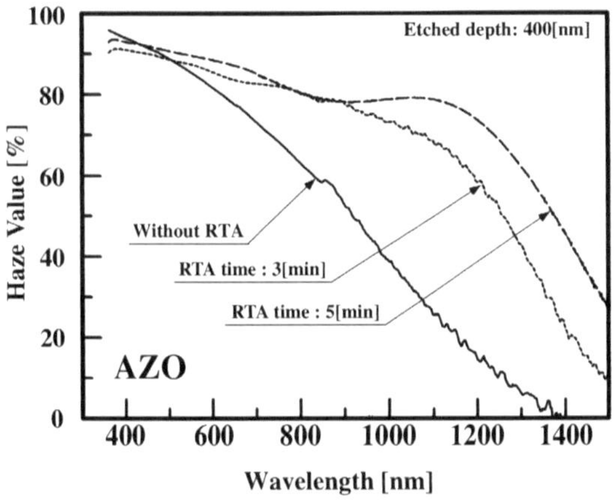

Fig.12 Haze spectra for surface-textured AZO films prepared by rf+dc-MSD and etched without RTA (solid line) and etched after being heat treated with RTA at 500°C for 3 (dotted line) and 5 min (dashed line).

Table 3 Electrical properties and etching rate obtained in as-deposited and heat-treated AZO thin films prepared by rf+dc-MSD.

	Before RTA (without RTA)	**After RTA (500[℃], 5[min], Air)**
Etching Rate [nm/sec]	**4.3**	**2.4**
Resistivity [Ω cm]	**3.06×10^{-4}**	**3.79×10^{-4}**
Carrier Concentration [cm^{-3}]	**4.89×10^{20}**	**3.45×10^{20}**
Hall Mobility [cm^2/Vs]	**41.7**	**47.8**

dc-MSD. In addition, a higher haze value and a broader haze spectra were obtained in AZO thin films heat treated with RTA at a temperature of 500 than 400°C as well as for 5 than 3 min.

Figure 12 shows the haze spectra of surface-textured AZO films that were etched without RTA treatment as well as after RTA treatment at a temperature of 500°C for 3 or 5 min in air. The etching was carried out using AZO films prepared with a thickness of 2000 nm by rf+dc-MSD and conducted to a depth of approximately 400 nm. The texture-etched AZO thin film that was etched after RTA treatment at 500°C for 5 min exhibited a higher haze value at longer wavelengths in the near-infrared region than an equivalent thin film etched after RTA treatment for 3 min: transmittance about 80 % and 60%, respectively, at a wavelength of 1200 nm. The improvements in haze value obtained in AZO films etched after heat treatment with the RTA treatment at a higher temperature are attributable to the increase in the size of the etch pits produced as well as the increase of transmittance in the near-infrared region that results from the plasma resonance frequency being decreased by the decreased carrier concentration, as shown in Table 3. The obtained electrical properties and etching rates (in a 0.1% HCl solution at 25°C) in an as-deposited AZO thin film prepared by rf+dc-MSD and those in an equivalent film after heat treatment with RTA at a temperature of 500°C for 5 min in air are summarized in Table 3. It should be noted that the etching rate of films was significantly decreased after heat treatment with RTA. The obtained haze spectra in surface-textured AZO thin films prepared by rf+dc-MSD and etched after heat treatment with RTA at 500°C for 5 min in air may be suitable for transparent electrode applications in Si-based thin-film solar cells fabricated with a micro-crystalline $Si_{1-X}Ge_X$ thin-film active layer that exhibits its maximum photo response at a wavelength near 1000 nm. It was also found that the obtainable haze value in GZO thin films prepared by dc-MSD and rf+dc-MSD was considerably improved by etching the films either before or after an RTA treatment. In contrast with the AZO films, the obtainable light scattering characteristics as well as the resulting surface-textured structure in texture-etched GZO film were relatively independent of the order that the

RTA treatment and the etching were conducted, irrespective of the type of deposition method used (detailed data not shown here). However, the obtainable haze value in GZO thin films was always lower than that in AZO films, when evaluated under equivalent conditions. A surface-textured GZO thin film that was prepared by rf+dc-MSD and etched after RTA treatment at 500°C for 5 min exhibited the highest transmittance, about 60% at a wavelength of 1000 nm. The obtained haze spectra in surface-textured GZO thin films prepared by MSDs and etched after the RTA treatment may be not appropriate for transparent electrode applications in Si-based thin-film solar cells fabricated with a micro-crystalline $Si_{1-X}Ge_X$ thin-film active layer.

3.2.2 Influence of the kind of doped impurity

The obtainable light scattering characteristics in impurity-doped ZnO thin films may be affected by the deposition method used. In the various deposition methods described above, the preparation of all the transparent conducting thin films, *i.e.*, AZO, BZO and GZO, is only possible with PLD; the preparation of BZO films by the MSDs and AZO films by VAPE was difficult because a sintered BZO target is not available and there is the large difference in vapor pressure between ZnO and Al_2O_3, respectively. In addition, a PLD method that utilizes plasma with an appropriate energy, as shown in Fig. 2, is suitable for the preparation of transparent conducting AZO, BZO and GZO thin films with lower resistivity, as seen in Tables 1 and 2 and Figs. 4-7, whereas the deposition rate is too low to use practically [26,29]. The light scattering characteristics were evaluated using AZO, BZO and GZO thin films all prepared with a thickness of approximately 2000 nm under equivalent deposition conditions by PLD to avoid the influence of the kind of deposition method used. It was found that the observed etch pit size from texture-etched films clearly increased when etched after heat treatment with RTA, irrespective of the kind of doped impurity. Figure 13 shows typical SEM images of the surface of surface-textured AZO, BZO and GZO films (to a depth of approximately 400 nm) that were etched after RTA treatment at a temperature of 500°C for 5 min in air. As can be seen in Fig. 13, the observed etch pit size from the AZO and BZO films always was larger than that from the GZO films. In particular, the etch pit size decreased in the following order: BZO, AZO and GZO thin films. The haze spectra are shown in Fig. 14 for the surface-textured AZO, BZO and GZO films that were etched after RTA treatment shown in Fig. 13. The difference in haze spectra among the texture-etched films doped with different kinds of impurity and heat treated with RTA is attributed to the difference in obtained etch pit sizes as well as the difference in transmission spectra, that may be related to the difference in obtained carrier concentration, *i.e.*, impurity content in the films, as seen in

(a) AZO **(b) BZO** **(c) GZO**

Fig. 13 SEM images of the surface of as-deposited and surface-textured (a) AZO, (b) BZO and (c) GZO films prepared by PLD and etched after being heat-treated with RTA.

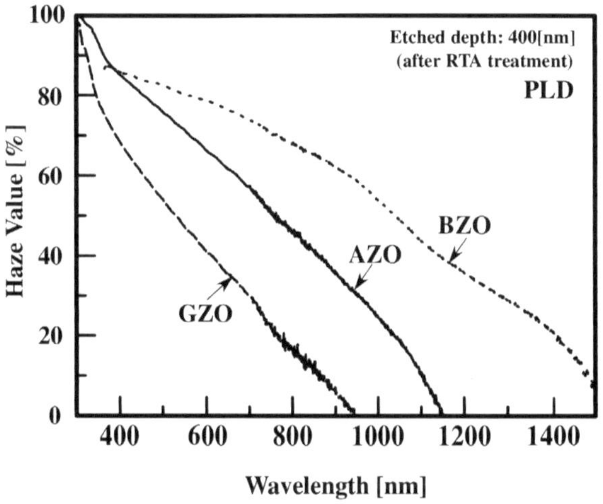

Fig.14 Haze spectra for surface-textured AZO (solid line), BZO (dotted line) and GZO (dashed line) films prepared by PLD and etched after being heat treated with RTA.

Table 2 and Fig. 7. It should be noted that the obtained haze spectra in AZO films prepared by rf+dc-MSD (by sputtering a sintered AZO target) and PLD (with the ablation of a part of the same AZO target used in the MSDs) and treated with the equivalent etching and RTA, as shown in Figs. 12 and 14, respectively, differ from each other. However, the AZO films prepared by rf+dc-MSD and PLD exhibited similar surface texture structures, as shown in Figs. 10(b) and 13(a), respectively. Thus, this difference in haze spectrum may be attributed mainly to a difference in obtainable carrier concentration rather than a difference in the crystallinity, such as c-axis orientation, resulting from the different deposition methods used, as seen in Tables 2 and 3. This suggests that evaluating the suitability of the light scattering characteristics in AZO, BZO and GZO thin films for Si-based thin-film solar cell applications may necessarily involve varying the content of the impurity doped into the films.

3. 3. The stability in moisture-resistance tests

Changes in the electrical properties of impurity-doped ZnO thin films were measured over time with exposure to a highly moist environment in air at 85% relative humidity and 85°C. We have reported that in long-term moisture-resistance tests, impurity-doped ZnO thin films exhibit an increase of resistivity that is considerably dependent on film thickness at levels below approximately 500 nm [8, 31, 50]. The resistivity as a function of test environment exposure time is shown in Fig. 15 for BZO (by VAPE) and AZO and GZO (by rf+dc-MSD) thin films prepared with a thickness of approximately 500 nm on glass substrates at 200°C. The content of B doped into the VAPE-BZO film was 1.2 at%, and the impurity contents doped into rf+dc-MSD-AZO and –GZO films are shown in Table 1. It should be noted that the stability of resistivity of BZO thin films in long-term moisture-resistance tests is lower than that of AZO and GZO thin films. In addition, it is known that most transparent electrodes used in thin-film solar cells in practical use require a thickness in the range from approximately 500 to 1500 nm as well as an ability to withstand exposure to a temperature of 85°C and a relative humidity of 85% [8, 31, 51, 52]. As can be seen in Fig. 15, the stability of VAPE-prepared BZO thin films with a thickness of about 500 nm in long-term moisture-resistance tests is insufficient for transparent electrode applications in thin-film solar cells. Since the long-term moisture-resistance tests showed that the stability of resistivity in VAPE-prepared BZO and rf+dc-MSD-prepared AZO and GZO thin films was dependent on the deposition method used, it should be noted that the obtainable resistivity stability in impurity-doped ZnO thin films may also be affected by the kind of doped impurity used. To avoid any influence from the deposition method used, the stability of resistivity in long-term moisture-resistance tests was evaluated using AZO, BZO and GZO films all prepared by the PLD method, as described above. Figure 16 shows resistivity as a function of test environment exposure time for BZO, AZO and GZO thin films prepared by PLD with a thickness of 500 nm on glass

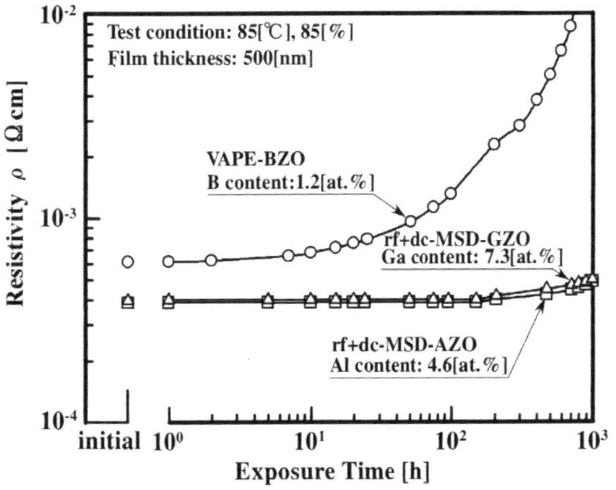

Fig.15 Resistivity as a function of exposure time (air environment: 85% relative humidity and 85°C) for BZO (○) (by VAPE) and AZO (□) and GZO (△) (by rf+dc-MSD) films prepared with a thickness of 500 nm.

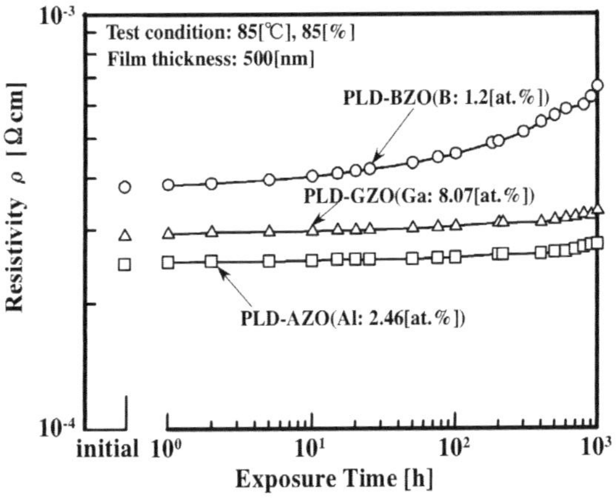

Fig.16 Resistivity as a function of exposure time (air environment: 85% relative humidity and 85°C) for BZO (○), AZO (□) and GZO (△) films prepared with a thickness of 500 nm by PLD.

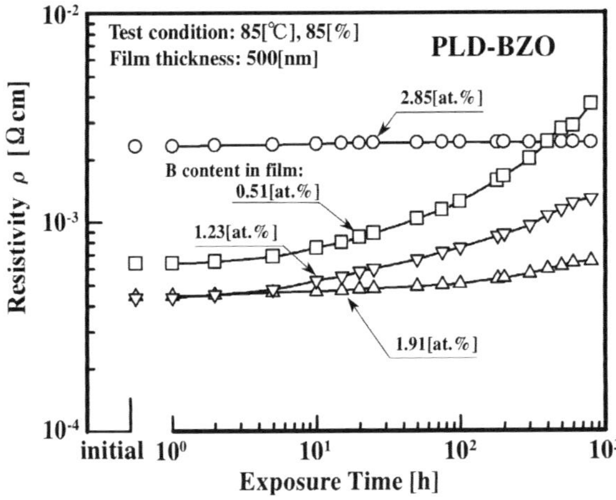

Fig.17 Resistivity as a function of exposure time (air environment: 85% relative humidity and 85°C) for BZO films prepared by PLD with a thickness of 500 nm and various B-doping contents: 0.51 (□), 1.23 (▽), 1.91 (△) and 2.85 (○) at.% by PLD.

substrates at 200°C, as shown in Table 1. Clearly, in BZO thin films prepared with a thickness of 500 nm by PLD, the long-term moisture-resistance test shows that the stability of resistivity is insufficient for practical use as transparent electrodes in thin-film solar cells. In contrast, the resistivity in the AZO and GZO thin films, also prepared with a thickness of approximately 500 nm, was stable, even after testing for 1000 h in a moist environment. In addition to the thickness dependence of the stability of resistivity in long-term moisture-resistance tests described above, we have reported that the resistivity stability in impurity-doped ZnO thin films is affected by the content of the impurity doped into the films [53, 54]. Examples of resistivity as a function of test environment exposure time are shown in Fig. 17 for BZO thin films prepared by PLD with a thickness of 500 nm and various B contents in the range from 0.51 to 2.85 at.%; the thin films exhibited an increase of resistivity that was considerably dependent on the B content. It should be noted that the stability of resistivity exhibited a tendency to improve as the B content was increased. The moisture-resistance tests of the BZO thin films showed that the increase in resistivity with exposure time can be attributed to decreases of both carrier concentration and Hall mobility, as also reported in AZO and GZO thin films prepared with a thickness below approximately 100 nm [53, 54]. This suggests that the resulting resistivity increase in BZO thin films is attributable to the grain boundary scattering brought about by decreases of both the carrier concentration and the Hall mobility that arise from the adsorption of oxygen on the grain boundary surface of polycrystalline BZO thin films, which is enhanced in a strongly oxidizing environment such as at higher relative humidities and temperatures [53, 54]. However, the obtained lower stability in BZO than in AZO and GZO thin films described above may be attributable mainly to the content than the kind of impurity doped into the films. Thus, the resistivity stability in BZO thin films prepared with a thickness above approximately 500 nm can be improved by increasing the B content doped into the films, which may cause a decrease in the Hall mobility and an increase in resistivity that depend on the level of B content doped, as seen in Fig. 4, irrespective of the deposition method used.

4. Conclusion

The development of transparent conducting impurity-doped ZnO thin films that would be appropriate for applications as transparent electrodes in thin-film solar cells was described. The suitability of the electrical properties and their stability as well as the suitability of the light scattering characteristics and the surface texture formation obtainable by wet-chemical etching were investigated in impurity-doped ZnO thin films prepared by various physical deposition methods. The suitability was evaluated using transparent conducting Al-, B- and Ga-doped ZnO (AZO, BZO and GZO) thin films prepared in a thickness range from 500 to 2000 nm on glass substrates at 200°C using pulsed laser deposition, vacuum arc plasma evaporation and different types of magnetron sputtering depositions. The evaluation of the suitability and the stability were focused, in particular, on AZO, BZO and GZO thin films prepared with an appropriate content of impurity doping that would result in the lowest attainable resistivity as well as films suitable for transparent electrode applications in flat panel displays. The obtainable Hall mobility in impurity-doped ZnO thin films with a resistivity on the order of 10^{-4} Ωcm was influenced by the impurity content doped into the films, *i.e.*, the obtained carrier concentration. The stability of resistivity of the BZO thin films in long-term moisture-resistance tests (in air at 85% relative humidity and 85°C) was found to be lower than that of the AZO and GZO thin films. The surface texture formation was carried out by wet-chemical etching (in a 0.1% HCl solution at 25°C) conducted before or after being heat-treated with rapid thermal annealing (RTA) or without RTA. In particular, a significant improvement of transmittance and haze value at wavelengths up to about 1200 nm in the near-infrared region that is attributable to a decrease of carrier concentration as well as an increase of etch pit size was obtained in surface-textured AZO films prepared by r.f. power superimposed d.c. magnetron sputtering deposition and etched after being heat treated with RTA at 500°C for 5 min in air. The obtained suitability and stability in impurity-doped ZnO thin films were related more to the content rather than the kind of impurity doped into the films as well as to the deposition method used. Further development of BZO and GZO thin films that would be appropriate for applications as transparent electrodes in thin-film solar cells will necessarily involve investigating thin films prepared by varying the content of impurity doped into the films.

Acknowledgment
The authors would like to thank T. Hirano, J. Nomoto and Y. Nishi for technical assistance in the experiments.

References

[1] Minami T, Miyata T 2008 *Thin Solid Films*, **517**, 1474.
[2] Yamamoto N, Makino H, Yamada T, Hirashima Y, Iwaoka H, Ito T, Ujihara A, Hokari H, Morita H, Yamamoto T 2010 *J. Electrochem. Soc.* **157**, J13.
[3] Tseng Z T, Kao P C, Chen Y C, Juang Y D, Kuo Y M, Chu S Y 2011 *J. Electrochem. Soc.* **158**, J310.
[4] Kim H, Horwitz J S, Kim W H, Makinen A J, Kafafi Z H, Chrisey D B 2002 *Thin Solid Films*, **420-421**, 539.
[5] Dawar A L, Jain A K, Jagadish C, Hartnagel H ed. 1995 "*Semiconducting Transparent Thin Films*", Inst of Physics Pub Inc.
[6] Minami T 2000 *MRS Bulletin*, **25**, 38.
[7] Minami T 2005 *Semicond. Sci. Technol.* **20**, S35.
[8] Nomoto J, Miyata T, Minami T 2011 *J. Vac. Sci. Technol.* A **29**, 041504.
[9] Wenas W, Yamada A, Konagai M, Takahashi K, 1991 *Jpn. J. Appl. Phys.* **30**, L441.
[10] Kushiya K, Ohshita Y, Hara I, Tanaka Y, Sang B, Nagoya Y, Tachiyuki M, Yamase O 2003 *Solar Energy Materials and Solar Cells*, **75**,171.
[11] Krc J, Zeman M, Kluth O, Smole F, Topic M 2003 *Thin Solid Films*, **426**, 296.
[12] Müller J, Rech B, Springe J, Vanecek M 2004 *Solar Energy*, **77**, 917.
[13] Yoo J, Lee J, Kim S, Yoon K, Park I J, Dhungel S K, Karunagaran B, Mangalaraj D, Yi J 2005 *Thin Solid Films*, **480-481**, 213.
[14] Beyer W, Hüpkes J, Stiebig H 2007 *Thin Solid Films*, **516**, 147.
[15] Rocjstuh C, Lederer F, Bittkau K, Carius R 2007 Appl. Phy. Lett. 91, 171104.
[16] Ruske F, Jacobs C, Sittinger V, Szyszka B, Werner W 2007 *Thin Solid Films*, **515**, 8695.
[17] Sittinger V, Ruske F, Werner W, Jacobs C, Szyszka B, Christie D J 2008 *Thin Solid Films*, **516**, 5847.
[18] Tohsophon T, Hüpkes J, Siekmann H, Rech B, Schultheis M, Sirikulrat N 2008 *Thin Solid Films*, **516**, 4628.
[19] Zhu H, Bunte E, Hüpkes J, Siekmann H, Huang S M 2009 *Thin Solid Films*, **517**, 3161.
[20] Kang D W, Kuk S H, Ji K S, Ahn S W, Han M K 2010 *Jpn. J. Appl. Phys.* **49**, 031101.
[21] Liu H, Avrutin V, Izyumskaya N, Özgür Ü, Morkoc H 2010 *Superlattices and Microstructures*, **48**, 458.
[22] Moon T, Yoon W, Ji K S, Ahn S W, Lee S, Joo M, Shin H Y, Park K, Lee H M 2010 *Appl Phy. Express*, 3 095801.
[23] Rath J K, Liu Y, Jong M M, Wild J, Schuttauf J A, Brinza M, Schropp R E 2010 *Thin Solid Films*, **518**, e129.
[24] Dewan R, Vasilev I, Jovanov V, Knipp D 2010 *J. Appl. Phys.* **110**, 013101.
[25] Krc J, Lipovsek B, Bokalic M, Campa A, Oyama T, Kambe M, Matsui T, Sai H, Kondo M, Topic M 2010 *Thin Solid Films*, **518**, 3054.
[26] Agra H, Suzuki A, Matsushita T, Aoki T, Okuda M 2003 *Thin Solid Films*, **445**, 263.
[27] Yamada T, Miyake A, Kishimoto S, Makino H, Yamamoto N, Yamamoto T 2007 *Appl. Phys. Lett.* **91**, 051915.
[28] Nomoto J, Hirano T, Miyata T, Minami T 2010 *Proc. of the 17th Int. Display Workshops*, 371.
[29] Tanaka H, Ihara K, Miyata T, Sato H, Minami T 2004 *J. Vac. Sci. Technol.* **A 22**, 1757.
[30] Miyata T, Honma Y, Minami T 2007 *J. Vac. Sci. Technol.* **A 25**, 1193.
[31] Nomoto J, Konagai M, Miyata T, Minami T 2010 *J. Vac. Sci. Technol.* **A 28,** 861.
[32] Nomoto J, Hirano T Miyata T, Minami T 2011 *Thin Solid Films*, online available October 8.
[33] Minami T, Hirano T, Noguchi Y, Miyata T, Nomoto J 2011 *Thin Solid Films*, to be published.
[34] Tominaga K, Yuasa T, Kume M, Tada O 1985 *Jpn. J. Appl. Phys.* **24,** 944.
[35] Tominaga K, Kuroda K, Tada O 1988 *Jpn. J. Appl. Phys.* **27,** 1176.
[36] Ellmer K, Cebulla R, Wendt R 1998 *Surface and Coatings Technology*, **98,** 1251.
[37] Ishibashi S, Higuchi Y, Ota Y and Nakamura K: *J. Vac. Sci. Technol.* A **8** (1990) 1403.

[38] Ichihara K, Inoue N, Okubo M and Yasuda N: *Thin Solid Films* **245** (1994) 152.
[39] Minami T, Miyata T, Yamamoto T, Toda H 2000 *J. Vac. Sci. Technol.* **A18**, 1584.
[40] Minami T, Sato H, Nanto H, Takata S 1985 *Jpn. J. Appl. Phys.* **24**, L781.
[41] Oda J, Nomoto J, Miyata T, Minami T 2010 *Thin Solid Films*, **519**, 2984.
[42] Minami T, Oda J, Nomoto J, Miyata T 2010 *J. Vac. Sci. Technol.* **A 28,** 842.
[43] Pisarkiewicz T, Zakrzewska K, Leja E 1989 *Thin Solid Films*, **174**, 217.
[44] Minami T, Sato H, Ohashi K, Tomofuji T, Takata S 1992 *J. of Crystal Growth*, **117**, 370.
[45] Minami T, Suzuki S, Miyata T 2010 *Mat. Res. Soc. Symp. Proc.* **666**, F1.3.1.
[46] Kluth O, Schöpe G, Rech B, Menner R, Oertel M, Orgassa K, Schock H.W., Thin Solid Films 502 (2006) 311.
[47] Sato H, Minami T, Takata S, Mouri T, Ogawa N, Thin Solid Films 220 (1992) 327.
[48] Nomoto J, Konagai M, Okada K, Ito T, Miyata T, Minami T, Thin Solid Films 518 (2010) 2937.
[49] Nagoya Y, Sang B, Fujiwara Y, Kushiya K, Yamase O, Solar Energy Mat. and Solar Cells, 75 (2003) 163.
[50] Minami T 2008 *Thin Solid Films*, **516**, 5822.
[51] International Electrotechnical Commission Environmental Testing-Damp Heat 60068-2-78.
[52] International Electrotechnical Commission Environmental Testing-Damp Heat 60068-2-3.
[53] Minami T, Miyata T, Ohtani Y, Kuboi T, *Phys. Stat. Sol.* (RRL), **1**, R31 (2007).
[54] Minami T 2008 *Thin Solid Films*, **516**, 1314.

IOP Conf. Series: Materials Science and Engineering **34** (2012) 012002 doi:10.1088/1757-899X/34/1/012002

Resonant Photoemission at the O1s threshold to characterize β-Ga_2O_3 single crystals

Marcel Michling and Dieter Schmeißer

BTU Cottbus, Applied Physics, Konrad-Wachsmann-Allee 17, 03046 Cottbus, Germany

dsch@tu-cottbus.de

Abstract. We report on spectroscopic investigations on β-Ga_2O_3 single crystals. We focus on the detailed analysis of the O1s resonance profile by resonant photoelectron spectroscopy. We analyze the electronic structure and assign the O2p to build the valence band and both, O2p and Ga4s4p states to contribute to the conduction band. We determine the partial density of states for the valence and conduction bands and find a strong hybridization of O2p and Ga4s states. This is deduced from constant final state spectra on the O-KLL-Auger over the O K-edge and Ga $L_{2,3}$-edge and a comparison to the corresponding X-ray absorption spectroscopy data. We also identify several types of defects. A broad band of oxygen derived defects is identified that extends throughout the gap. Small polarons are attributed to cause an anti-resonance in the valence states around the O1s threshold. In addition, a separate Auger decay at resonance indicates the existence of localized charge transfer states which involves localized Ga4sp states.

1. Introduction

β-Ga_2O_3 (100) is a transparent conductive oxide (TCO). Single crystals of this material are available which enable defined spectroscopic investigations of in UHV-cleaved samples. Recent studies had focused on the structural and transport properties [1], low energy electron diffraction (LEED) and scanning tunnel microscopy (STM) [2], and valence band (VB) angle resolved photoelectron spectroscopy (ARPES) measurements [2, 3]. A recent report compares the valence band data of In_2O_3 and Ga_2O_3 [4]. Related studies deal with Ga_2O_3 powders and Sn-doped Ga_2O_3 samples [5, 9].

The general challenge in TCOs is to understand how in a material which has a broad band gap - in order to become optical transparent – simultaneously a high conductivity can be achieved. Certainly, this challenge addresses the existence of defect states but again these should be transparent also. Of fundamental interest in basic research therefore is to learn about the size and character of the band gap as well as about the possible defect mechanisms.

β-Ga_2O_3 crystallizes in the base-centered-monoclinic structure (space group C2/m) with lattice parameters a =1.22140nm, b = 0.30371nm, c=0.57981nm, β = 103.83°, and 4 Ga_2O_3 f.u./unit cell [6]. Ga is in the same row of the periodic table as Al and In. It has the atomic

Published under licence by IOP Publishing Ltd

configuration $Ga3d^{10}4s^{2}4p^{1}$. In the oxide Ga is three-valent with the $Ga3d^{10}4s^{0}4p^{0}$ configuration. In the unit cell of β-Ga_2O_3 there are two different Ga sites as well as three different oxygen sites [4, 7]. Consequently, the electronic structure is determined by the O2p states in the VB and the Ga4s4p states which form the conduction band (CB) [8].

In theoretical studies the basic features of the electronic structure of both, the α- an β-phases of Ga_2O_3 like band structure, partial and joint density of states (DOS) are reported [8] and a strong covalent bonding is concluded [8] and [9]. All theoretical studies [4, 8, 9] agree on the partial DOS (pDOS) for the VB is predominantly of O2p character. The conduction band states are predominately derived from Ga4sp states mixed with some contributions of O2p. In a recent communication it is reported that there is evidence in electron paramagnetic resonance (EPR) data for two different defect sites [1]. A shallow donor has been identified as well as a localized defect site which is seen in the EPR data.

2. Experimental

β-Ga_2O_3 single crystals [1] are mounted on our transfer system and cleaved in UHV. We investigated in total three different crystals which had different colors (from transparent to yellow) because of the amount of incorporated carbon. We found no noticeable differences in the VB and core level data for these samples. The spectroscopic measurements were carried out at the U49/2-PGM2 at BESSY II in Berlin [10]. We apply the methods of photoelectron spectroscopy (PES), resonant photoelectron spectroscopy (resPES) and x-ray absorption spectroscopy (XAS) [10]. Constant final state (CFS) data are recorded at a fixed kinetic energy; constant initial state (CIS) data are taken recorded at affixed binding energy of the electron energy analyzer.

3. Results

Here we first report on the photoemission data and later concentrate on the XAS spectra at the O1s and Ga2p threshold.

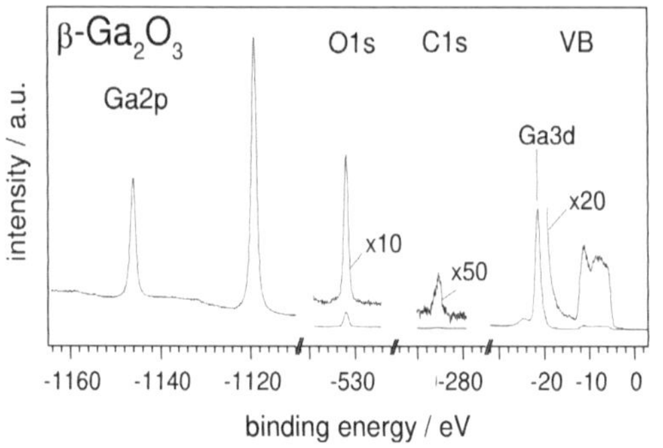

Figure 1: The x-ray photoelectron spectroscopy (XPS) data of β-Ga_2O_3 as recorded at 1300eV excitation, shown are the ranges for the Ga2p, O1s, C1s core levels as well as the valence band regime with the Ga3d and O2s emissions.

In figure 1 we show the photoemission data of the UHV cleaved surface excited by 1300eV photons. The core levels of the Ga2p states appear at 1119.3eV binding energy (b.e.) and show a very sharp structure with a full width of half maximum (FWHM) of 1.35eV. There is a weak satellite feature which appears about 13eV from the main lines.

Traces of Carbon are observed in the C1s data which most probably come from CO_2 which has been incorporated by the growth process.

The O1s level appears at 532.0eV and has a single Gaussian peak structure with a FWHM of 1.1eV. This finding is remarkable as in most other oxidic systems in the O1s signal there is shoulder at 533.3eV. The missing of this shoulder underlines the high quality of the Ga_2O_3 single crystals and the surface preparation by UHV cleavage.

Very important for the discussion of the electronic structure is the valence band regime which is dominated by the shallow core level of the filled $Ga3d^{10}$ states at -21.7eV. In figure 2 we show the valence band photoemission data now excited by photons of 150eV. We identify three pronounced maxima which we can attribute to the three different oxygen sites (O^1, O^2, O^3 refer to figure 8) expected in the Ga structure.

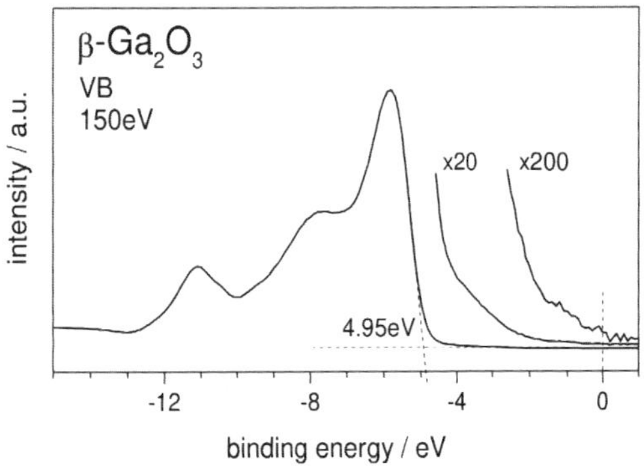

Figure 2: The valence band photoemission spectrum of β-Ga_2O_3 as recorded with 150eV excitation. The position of the valence band maximum is indicated. The enlarged parts visualize the weak emissions within the band gap regime.

Here we also focus on the position of the valence band maximum (VBM) which can be easily extrapolated and is found at 4.95eV. The total width of the valence band photoemission is about 8eV, in reasonable agreement with earlier reports. It appears to be somewhat larger than a calculated band structure of 7.26eV [3], an experimental value of 7.37eV [3] and another calculated value of 7.00eV [8]. Blowing up the emission above the VBM, we find a strong decaying emission which becomes very weak but still noticeable up to the Fermi energy. This is a clear evidence for gap states which will be discussed in more detail below. This observation is remarkable as we use high quality n-type single crystals [1] showing carrier densities of as low as 10^{17} cm^{-3}.

In figure 3 we show again the valence band regime as excited by 1300eV photons (figure 1) but now we focus on the experimentally observed loss features at the high binding energy side of the Ga3d level (x14). This loss feature shows up also at the O1s core levels (not included). The Ga3d core level is an ideal primary peak because of its low width. It appears very sharp for β-Ga_2O_3 thereby enabling a founded analysis of these weak signals. We have added the calculated loss function [8] as well as the experimental data for Ga_2O_3 nanoparticles by referring to a common loss energy scale. To obtain a better agreement to our data we applied a rigid shift by 1.37eV and 4eV, respectively. The agreement is very good as a general shift of the loss function just is attributed to problems in determining the band gap. It indicates that the calculated band structure gives a good description of the electronic properties in VB and CB regime. The larger broadening of the experimental data can be associated with defects as the loss function does not include any defects.

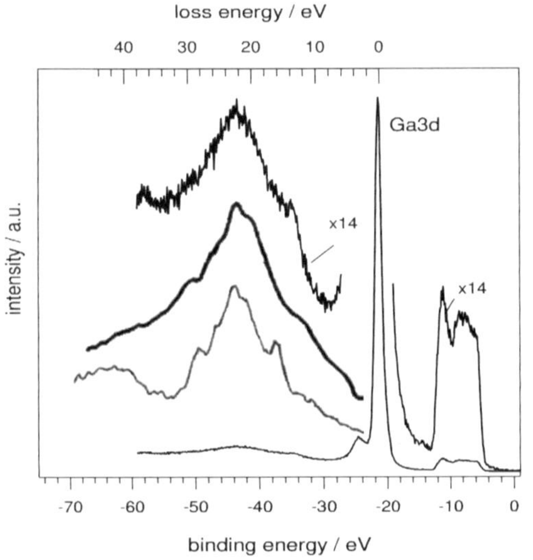

Figure 3: The Ga3d emission of β-Ga_2O_3 as recorded with 1300eV photon energy. In enlarged (x14) scale the loss features at the Ga3d (middle) are shown. For comparison we give also the experimental data for Ga_2O_3 nano-particles (bold line) and the calculated loss function for the β- phases of Ga_2O_3 [8, 11] (lower curve).

We now focus on the XAS data. In figure 4 we show the XAS-TEY (total electron yield) curve for the β-Ga_2O_3 crystal. There are two broad absorption bands which peak at 535.5eV and at 541eV. The absorption starts at around 532eV and there is a weak shoulder in the rise of the first absorption band. Its position is at 533.2eV, it is marked by an asterisk.

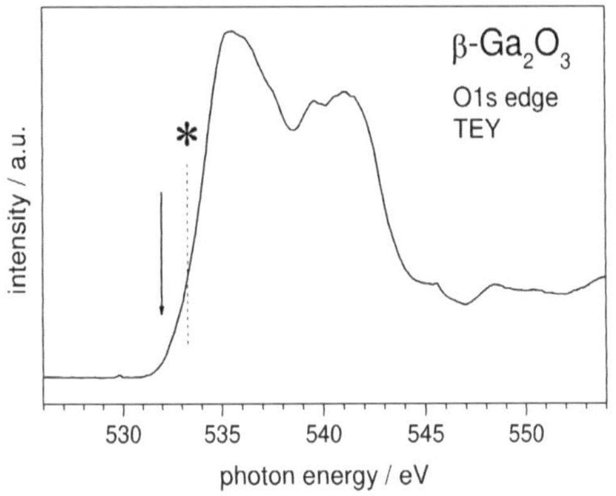

Figure 4: The X-ray absorption spectrum of β-Ga_2O_3 at the O1s threshold as recorded in the TEY mode. The arrow marks the position of the Fermi energy; the asterisk indicates the onset of the 3h-Auger process (see text).

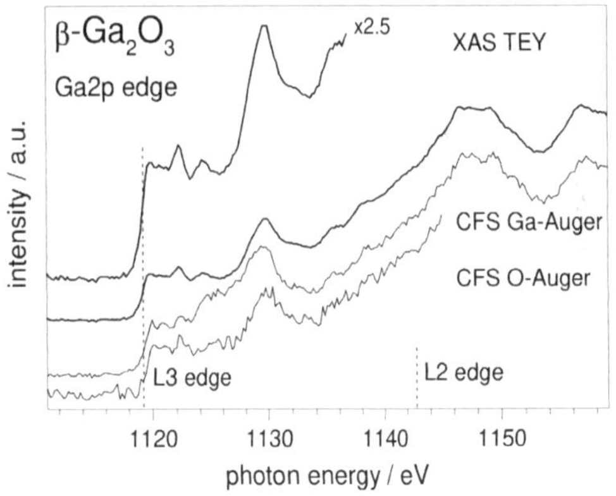

Figure 5: XAS data taken at the Ga2p-edge. The TEY spectrum is shown as measured and enlarged (x2.5). It is compared to the CFS spectra which are recorded at the kinetic energies of the Ga-Auger and of the O-Auger transitions.

In figure 5 we show the XAS data taken at the Ga2p edge. In the figure we included the TEY spectra as well as the CFS data as recorded at the O-KLL Auger (E_{kin} = 512eV) and at the Ga-LMM Auger (E_{kin} = 777eV). In general these signals are rather weak, since transitions from Ga2p into the CB states should involve only Ga4s states. The intensity of such transitions depends on the number of empty sites and is small as expected hence. The TEY signal shows a steep rise at 1119.25eV photon energy followed by a plateau which has three main features at 1119.9eV, 1122.2eV and at 1124.2eV. In such the spectrum resembles the corresponding data of other $3d^{10}$ systems like Cu and Zn. The two CFS data displayed are weaker in intensity and exhibit a higher noise level but in total show a very similar profile as all the three curves at the L_3 edge start at 1119.25eV.

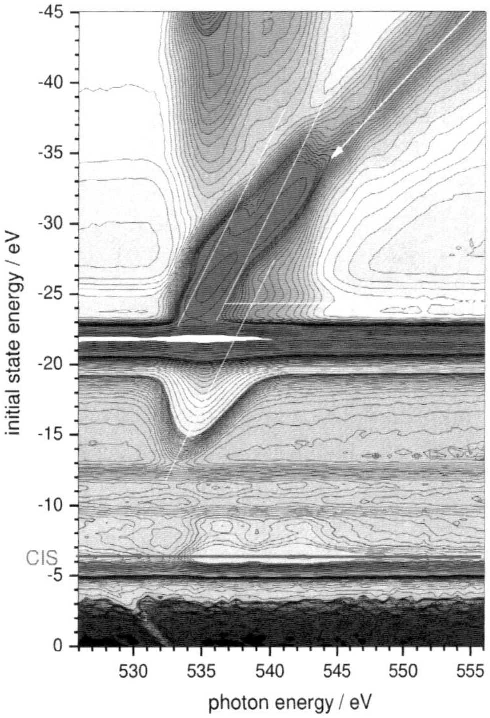

Figure 6: resPES data of β-Ga_2O_3 at the O1s absorption edge. The arrow indicates the 45° Auger and the lines mark the 67.5° Auger processes. The red line marks the CIS spectrum shown in figure 7.

Around the O1s threshold we record a complete set of energy distribution curves (EDC) to further characterize the oxygen derived states. In res-PES an electron out of the O1s core level is resonantly transferred into the empty CB states. It causes an excited intermediate state in which only oxygen derived empty states are involved due to the dipole selection rules. The decay of these states can be used to probe the empty CB states. The plot in figure 6 was acquired by variation of the photon energy in the range between 526eV and 556eV. In this contour plot of the valence band spectra recorded around the O1s threshold we can distinguish between features at constant binding (horizontal) or constant kinetic energy (45°). Here we just focus on the O-KLL Auger lines with a kinetic energy of 512eV, which become evident in the profile as lines following an angle of 45°. Such Auger contributions are observed for photon energies above 547eV and are indicated in figure 6 by the white arrow. However, it is evident in that plot that the features next to the threshold at around 531.5eV are stronger inclined and appear under an angle of 67.5°. These are indicated by white lines in the regime between 530eV up to 540eV. In addition we have marked by horizontal lines a participator decay (white line) as well as the position of the CIS data shown in figure 7 (black bold line).

The observation of the 67.5° Auger lines indicates that there must be either different mechanisms or different states involved in the resonant behavior at the O1s threshold. We argue that in this regime we have contributions from another Auger process which has three holes in its final state and will focus on that process in our discussions below.

IOP Conf. Series: Materials Science and Engineering **34** (2012) 012002 doi:10.1088/1757-899X/34/1/012002

In figure 7 we focus on the CIS spectrum taken on the most prominent valence band peak at a binding energy of -6.25eV (O^1 in figure 8). This spectrum shows two main observations. First, there is a clear anti-resonance which starts already at 522eV. Second, there are pronounced resonances peaking at 535.5eV and at 541eV. The latter two features are consistent with those of the TEY data shown in figure 4. We will discuss the two observations in terms of oxygen based defects and strong covalent bonding, respectively.

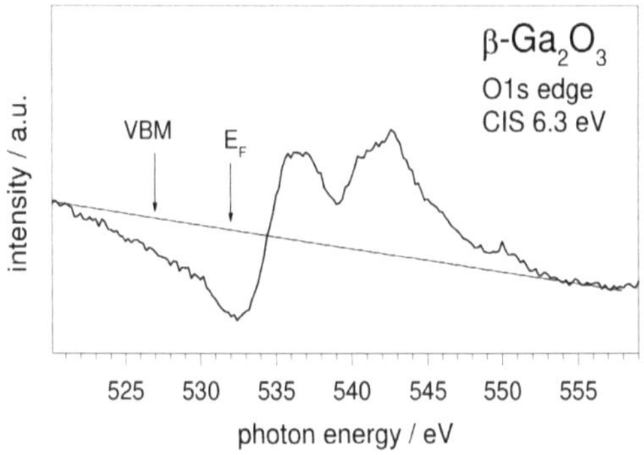

Figure 7: The CIS data of the valence feature O^I at the O1s threshold. The positions of the VBM and of the E_F are marked by arrows.

Next, we give a combined electronic structure of the respective O1s and Ga2p XAS and VB data at resonance (figure 8 upper panel). To obtain this figure we have used the binding energies of the O1s (532eV) and Ga2p (1119.3eV) core levels (figure 1) and referred the XAS data accordingly to obtain a common energy scale with respect to E_F. The VB spectrum is taken at 150eV (figure 2) excitation energy. Within this diagram the pDOS can be identified via the resonant behavior of the oxygen and gallium derived states. The O1s-TEY shows a slow increase at the Fermi energy while the step of the Ga2p states appears right at E_F. These combined data enable us to discuss the electronic structure in more detail. The combined UV photoelectron spectroscopy (UPS) and XAS data can be related to the oxygen and gallium derived partial densities of states, respectively. From these data, we deduce directly the electronic structure of the VB and of the CB states, hence.

Information about defect states will be deduced from the CIS data which are displayed in the lower panel of figure 8. Here it becomes evident that the anti-resonance extends not only throughout the band gap but starts already at energies which correspond to the middle of the valence band. The CIS data which sample the onset of the 3h-Auger (CIS 13.75eV) is also included. Its maximum coincides with the shoulder in the O1s TEY spectrum (asterisk in figure 4). This feature is attributed to represent a charge-transfer state from an oxygen ($O2p^5$) ligand to a localized Ga4sp state.

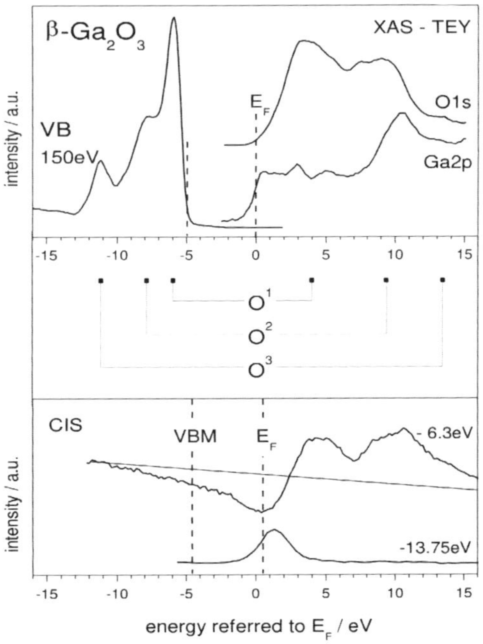

Figure 8: The combined partial density of states of the valence band and conduction band as determined from the UPS and XAS data, respectively (top). The contributions of the three different oxygen sites O^1, O^2, and O^3 are indicated in the center of the panel (see text). In the lower part we compare to the CIS data of figure 7 and show the CIS spectrum (-13.75eV) at the onset of the Auger process at the O1s resonance. The common energy is referred to E_F. The dashed lines indicate the position of the VBM and the Fermi energy.

4. Discussion

In this section we will first discuss the band structure of Ga_2O_3 and the relative contribution of oxygen and Ga states within the CB. Next we focus on intrinsic defect mechanisms which lead to the existence of band gap states. We identify these states from two novel mechanisms: We report on the Fano anti-resonance which we attribute to a band of oxygen derived hole states. Finally we discuss the 3h-Auger mechanism at the O1s threshold as a new process to identify charge transfer (CT) complexes in Ga_2O_3.

The availability of good quality single crystals of β-Ga_2O_3 enables [1] a systematic investigation of the electronic structure. For the detailed analysis of valence and band gap states this is an essential prerequisite as otherwise the origin of intrinsic electronic defect mechanisms might be blurred by some influence of structural defect sites. In particular, the often quoted argument of "some influence of grain boundaries and defects of all kind" does not hold anymore as such defects can be excluded definitely.

In this context we like to focus on the observation that the O1s core level in our data (figure 1) consists of a single Gaussian line. This contrasts the finding in most other oxidic systems where the main line is followed by a shoulder separated by about 1.5eV in energy. While the discussions about its origin are still ongoing we explicitly can rule out the presence of residual OH groups or dumbbell di-oxygen species. On our cleaved samples we definitely do not have such contributions at all. Comparing our data to those of a recent study on In_2O_3 single crystals we can also argue that in our Ga_2O_3 samples the number of oxygen vacancies must be below the detection limit of core level spectroscopy.

These findings at the O1s core level underline the superior quality of our Ga_2O_3 single crystals. A consistent interpretation of the electronic structure including all of the experimental data becomes possible, hence.

Following a general approach [12] for oxidic semiconductors in Ga_2O_3 the valence band states arise from the O2p states while the conduction band states arise from Ga4s and Ga4p states. This covalent bonding interaction can be easily understood as the symmetry of the Ga4s states enables an easy bonding with the p-states of the oxygen – at least much more easy than the bonding of the complex orbitals of 3d states in transition metal wave-functions. In the detailed band structure calculations this assignment is confirmed in general [5-8] however there is only a very weak covalent contribution as the main bonding character is found to be due to ionic interactions.

A direct comparison to the calculated electronic structure can be obtained via the dielectric loss function. The loss function describes all optical transitions from occupied VB states into empty CB states. It represents the joint DOS (JDOS) as optical transitions only occur vertical at $\Delta k = 0$, the loss function is the sum over all possible optical transitions. The JDOS is a crucial test for theoretical band structure models and also not easy to determine experimentally. In most oxide systems the O1s core level is used to derive the loss function from the electrons scattered on the higher binding energy side of the peak. However, the O1s core level in most systems is broader than 1.1eV and has shoulders as discussed above. Here, in β-Ga_2O_3 we can use the sharp Ga3d core level which has a high intensity and a small FWHM. We find that the observed loss function is in reasonable agreement with the calculated function [8]. Also the agreement with the experimentally derived loss function on Ga_2O_3 nano-particles [8] is almost perfect (see figure 3). Insofar, also the theoretically deduced joint DOS loss function gives a rather good agreement to our data (see figure 3 and 5) regarding the main features of the band structure. However, it should be mentioned that the loss function is not sensitive for defect states.

4.1 covalent contributions in the electronic structure of β-Ga_2O_3

Now we focus more in detail on the covalent contributions in the electronic structure of β-Ga_2O_3 which are reported also in the literature [5-8]. These are considered to be weak as deduced from the amount of pDOS of oxygen within the CB regime. In our data we derived the pDOS directly from the XAS data shown in figures 4, 5 and8 and we conclude that the covalent strength must be much stronger than anticipated from theory so far.

Another detail concerns the distribution of the O2p states within the conduction band. In our data in figure 2 we find that the valence band PES data shows three well distinguishable features. In a first attempt we may assign these bands with the structural in-equivalent oxygen sites known from the unit cell. Accordingly, the strongest contributions in the first band below the VBM can be associated with the oxygen atoms at the tetrahedral Ga sites (b.e. = 6eV, O^1). In more details, these in-equivalent sites of the unit cells cause a different bonding strength for the individual oxygen atoms a finding which is in agreement with the calculations [8, 9]. In addition our data show that this bonding strength is not uniform for the three different oxygen sites. We suggest that the sequence of the three in-equivalent oxygen atoms can be assigned as posted in our data (figure 8). In the VB spectra the three bands are assigned in the sequence O^3, O^2, and O^1. Also in the CB this assignment can be done, referred to the O1s-XAS data in the sequence is O^1 (536.5eV), O^2 (541eV), and O^3 (545eV).

The oxygen atoms with most pronounced covalent contributions are the O^1 states. The covalent admixing is witnessed in the corresponding XAS data by the strong resonances at 536.5eV and 541eV. These are a direct measure of the pDOS of states in the CB and visualize the degree of covalence for these states, hence. The strongest bonded oxygen atoms are those at the octahedral Ga sites (b.e. of -11eV, O^3). These have a much smaller covalent contribution as they do not contribute to the pDOS in the CB.

Also the features of the constant-initial-state data reflect the pDOS in the CB. There are pronounced intensity modulations of the valence band states (CIS -6.25eV in figures 5 and 7) when sweeping over the O1s threshold. These data represent even stronger evidence and a measure for the covalent bonding strength in oxidic systems as they link the O2p VB states directly with the empty CB states. The broad humps with maxima at 536.5eV and 541eV describe the admixture of empty oxygen states into the CB states. On the other hand, the contribution of the O^3 site at 545eV is weaker while O^2 has a noticeable contribution in the CIS profiles. These intensity modulations are also noticeable within the valence band regime of the resPES data shown in figure 6.

For the following discussion we have referred the XAS data to the Fermi energy using the value of the binding energy of the O1s level. Actually this enables us to mark the position of EF and VBM within the CIS data in figure 7. We notice two remarkable findings. First, the covalent contributions of O2p derived (O^1) pDOS starts at about 2eV above the Fermi energy, second, the anti-resonance behavior starts already below the VBM.

In the same manner we have combined the valence band data and the TEY-XAS and CIS data at -6, -13.75eV in a diagram with a common energy scale. Also in this combined diagram it becomes evident that the oxygen states do not contribute to the lowest CB states. We find that in the TEY-XAS data there is a weak onset starting at EF but the maximum is found almost 4eV above EF.

In contrast, the Ga-derived pDOS starts readily at the Fermi energy. To obtain this diagram we have used the value of the binding energy of the Ga2p core levels. We find that in the Ga-derived pDOS there must be a strong admixture of the Ga4p states as otherwise we would not observe the strong intensity in the Ga2p XAS spectrum. This observation is in contrast to the calculated pDOS [5] and also not in agreement with the band structure calculation where the lowest band in the CB regime shows a parabolic dispersion which is characteristic of pure Ga4s-derived states.

To summarize this part of the discussions we a good agreement of the main features in the electronic structure with calculated band structure regarding the JDOS. However, regarding the O2p pDOS in the CB we have to admit that there are much stronger covalent contributions than predicted from theory. Also, the significant contribution of Ga derived states next to E_F is a finding not predicted by theory.

4.2 Intrinsic defects

The next part in our discussion will deal with the contribution of intrinsic defects to the electronic structure. In our data we also have evidence for intrinsic defects – which are not treated in the calculations so far. Evidence for defects is already given in figure 2 where in the valence band photoemission spectrum in the regime above the VBM we find an almost continuous distribution of gap states (see figure 2). We observe a rather strong contribution which extends up to 1eV above the VBM (x20 in figure 2). These defect states show a tail which reaches up to 3eV above VBM. Here the defect intensity becomes very weak but is still noticeable, indeed until it reaches the Fermi energy (x200 in figure 2). Usually such features are attributed to shallow defect states; however, "shallow" refers to states which extend only some 100meV into the gap. This is clearly not the case in these Ga_2O_3 samples. Therefore we argue that the photoemission within the gap indicates some more extended defect states.

In fact, we observe two defect mechanisms in the data at the O1s threshold of oxides and which will become of significance here for the discussion of electronic defects in β-Ga_2O_3.

4.2.1 Small polarons – oxygen hole continuum

The first defect state is attributed to the observation of a Fano profile in the main peak which shows up in the valence band spectra at -6eV. This photoemission intensity shows a pronounced anti-resonance when passing over the O1s threshold in resonant excitation. The profile in particular is recorded in a CIS spectrum and shown in figures 7 and 8. The anti-resonance starts at around 522eV and extends up to 540eV. An anti-resonance associated with a Fano process is explained [14, 15] by an interference process of two similar photoelectron wave functions. Fano has argued [15] that the interference between two different photoemission channels can cause an asymmetric line profile and introduced a parameter $q = ß^2/\alpha^2 * V$ which gives the ratio of the squared amplitudes of the two different wave functions, V is the coupling between the two channels. An anti-resonance is obtained for q values below 1, a q parameter of about 0 means that the two channels do exist but the amplitude of the main channel α is much higher than that of the localized wave function β.

In order to explain the atomic origin of an anti-resonance at the O1s threshold we only can consider oxygen derived defect states because of the dipole selection rules which have to be applied for the resonant excitation out of the O1s core level. We argue that these transitions must involve $O2p^5$ defects, hence, as these have an empty site available for the incoming electrons. In the decay process we then attribute the indirect channel to proceed via an $O2p^{5+1}$ resonant state to cause an $O2p^4$ final state via an auto-ionization process. The direct channel is the photoemission process out of the $O2p^5$ states. Both channels end with the same $O2p^4$ final state. This can be understood that the escaping photoelectron has the chance to "visit" the empty site in an $O2p^5$ state. The time used for that visit gives rise to a phase shift causing the interference with the normal photoemission process. In other words, the q=0 Fano profile is an indication of the existence of $O2p^5$ initial states as otherwise a second channel via the $O2p^6$ intermediate state would not be possible at all.

In our data we do not have a single Fano profile but rather a very broad distribution of individual Fano processes. Their distribution is continuous over the electronic band gap with an increasing intensity towards the Fermi energy. It should be mentioned that the onset of these anti-resonances starts already at 522eV. In figure 7 we have added the arrows which mark the position of the VBM and the Fermi energy. It is evident, that the photoemission already in the valence band states O^1 and O^2 (see figure 8) contributes to the Fano profile. This means that the uppermost states in the valence band must have at least some contributions of $O2p^5$ defects. In other oxide systems where the covalent contribution are less pronounced we find that the $O2p^5$ states extend over the complete width of the CB also.

Remarkable is that only this first peak of the VB photoemission data shows the anti-resonant behavior while the other VB features are clearly less involved. It is the O^1 (tetragonal coordinated oxygen) peak which is that with the weakest covalent contribution. One idea could be that these states are more easily converted into defect states as any local change can reduce the covalent splitting further resulting in a gap state, hence. In contrast, the octahedral coordinated oxygen (O^3) is assumed to have the highest binding energy. This is in agreement with the band dispersion in the ARPES data [3, 4] as these bands show the largest dispersion in the valence band regime.

We attribute that these states originate from the small polaronic states and now discuss the main features "small polarons" [16]. This kind of defect is stabilized by various contributions. In general, there are lattice distortions, variations in the local dielectric properties, structural distortions among others [16]. All of these effects contribute and cause an energetic contribution which leads to a stabilization of a hole on the oxygen site. This hole state is localized and weakly correlated by that extra energy which means that these hole states cannot be treated within Fermi statistics and not be attributed to a normal "doping" procedure.

The most important phenomenon is that the oxygen atoms in the oxygen sub-lattice generate their own hole transport, the mobility of the small polarons is attributed to a charge transport by hopping. The negative charge then can be treated separately. It may be localized at the metal site causing a local reduction and change in the metal's oxidation state or may build a negative polaronic state. Therefore the corresponding M^- polarons are expected generally where electrons are donated to the metal sub-lattice.

Another important property of small polarons is that their existence implies that these polaronic features are insensitive to the identity of their originating lattice distortion. It might be that different dopants cause similar deformations which stabilize the formation of small polarons.

4.2.2 3h Auger decay - CT transitions

The second defect state we associate with the CIS (-13.75eV) spectrum as shown in figure 8 and which shows a peak right at the O1s resonance. In the resPES diagram shown in figure 6 we find that the first absorption band of the oxygen resonance is due to an Auger process. This process, however, must be different from the common spectator decay which at resonance in our data would occur under 45°, instead – as is evident from figure 6 – it precedes under 67.5°. This value of inclination differs from the usual Auger decay mechanism which appears under 45° and shows up in the resonances above 534eV. We explain such a behavior by the 3h-Auger process and we adopt here an explanation which was first given for an organic semiconductor [17]. Accordingly, such an Auger decay is attributed to a "delayed" participator decay in which the participating electron is parked in an localized defect band gap state. As a consequence the life time of the electron is longer than that of the primary core hole. The latter is filled by some valence electrons during the life time of the excited electron state. The participator process can be completed and the excited electron is emitted only by delivering another core hole. As a result there are three valence holes involved in the process, giving rise to the novel decay channel as shown in the resonance profile of figure 6.

A longer life time of the excited state is attributed to the resonant excited electron is no longer within the conduction band but is trapped into a localized gap state. In the organic semiconducting system this state is an excitonic state within the gap. The close correspondence to the spectroscopic 67.5° decay leads us to conclude that such localized states must be involved in this oxidic semiconductor also. From the appearance of the peak in the CIS (-13.75eV) spectrum in figure 7 we learn that these states appear in energy just around the Fermi level.

The 3h-Auger process is a final state process in which the primary core hole at the O1s level is filled to cause three valence holes V^1, V^2, V^3, and the resulting kinetic energy is transferred to the resonantly excited electron to leave as participator decay. The resulting final state may be denoted as (O1s – V^1 – V^2 – V^3 – P), hence. The first step is the excitation into the empty CB from the O1s level which causes the creation of the primary core hole. This primary electron becomes trapped into an empty localized state. Here the electron is stabilized by a Coulomb energy, which means that it has a longer lifetime when compared to a conduction band state. Consequently, while it is in the localized state, its primary core hole can be filled from any valence band state. This creates two valence holes (V^1 – V^2) by a common Auger decay. Up to this point, the process is just typical spectator decay. The three hole process now involves the primary electron in its localized state to become emitted as a participator Auger electron. This step creates the third valence hole V^3, the electron of which is to fill an additional, accidently bypassing core hole and transferring its energy to the primary electron to escape. As a result, we have to deal with a combined spectator – participator mechanism that is enabled by two facts, first it is important to have a localized state available and second it requires that the valence holes are delocalized covalent bonded O2p states. Furthermore, this mechanism becomes possible only at resonance where a high density of primary core holes is available. In addition it requires a high mobility of valence band electrons and holes, which is given only in covalent bands. These arguments make clear that the 3h-Auger process is not an atomic-like process (as the KLL Auger decay is) but is a multiple particle decay channel at resonance.

The main feature of the 3h-Auger process in the context of the electronic structure of Ga oxides, however, is that the states involved must be empty or at least only partially filled in order to accommodate the primary electron. Further, they need to have some Coulomb interactions U in order to trap the primary electrons. Consequently, the observation of the 3h-Auger is a tool for identifying such partially filled defect states. Insofar, we can use the observation and evidence of the 3h-Auger process as a method to probe the existence of correlated defect states within the probed system. The 3h-Auger process proceeds via localized empty states within the electronic band gap. In Ga_2O_3 such states are attributed to the oxygen derived states which are involved in a charge transfer process in which an electron from the O2p valence states is transferred into an empty metal state. Such processes are usually observed in the transition metal oxides in which the 3d occupation is increased thereby ($3d^{n+1}L$, $O2p^5$=L). In fact, in such system we do observe the 3h-Auger process [18] and it is as pronounced as we observe it here.

For Ga_2O_3 the problem with CT states is that they are known only for 3d-TM and not for closed shell oxides. One possible explanation is that the $3d^{10}$ configuration is not really completed but has some exchange to the 4s and the 4p states resulting in a $3d^{10-\delta}$ $(4s4p)^{+\delta}$ configuration. This scenario is indeed found in In_2O_3 [4] and shows up in a shoulder at the In3d core level. For β-Ga_2O_3 we do observe very sharp Ga2p core levels (figure 1) with no indication of such charge reorganization phenomena. Hence, this argument is most probable not valid. We rather favor the idea that the CT requires a full charge transfer and is not possible with partial charge exchange as in covalent systems. This is a consequence of the correlated nature of the CT mechanism. On the other hand, a pure 4s4p charge transfer is not unlikely at all. Although we are expecting that the 4s band is mostly parabolic, already any admixture of 4p states will create flat bands which can be the origin of charge localization. Such flat bands may be different for the three in-equivalent oxygen atoms and may cause multiple CT levels, hence.

Another argument in favor of the CT states at the Ga4sp levels is in the observation, that the pDOS for the Ga derived CB states starts right at the Fermi energy. This is not expected by the calculations [8, 9] in which the Ga4s contributions appear several eV above those of the O2p derived states. In our data this is reversed. As an explanation we attribute the Coulomb interaction U_{vh} between the O2p valence hole and the metal electron state to the fact that the Ga states shift down in energy to occur next to E_F. In other words these Ga derived states next to E_F give direct evidence for the CT process. Hence, our data are unambiguously convincing and confirm that CT into Ga4sp states really does exist in Ga_2O_3.

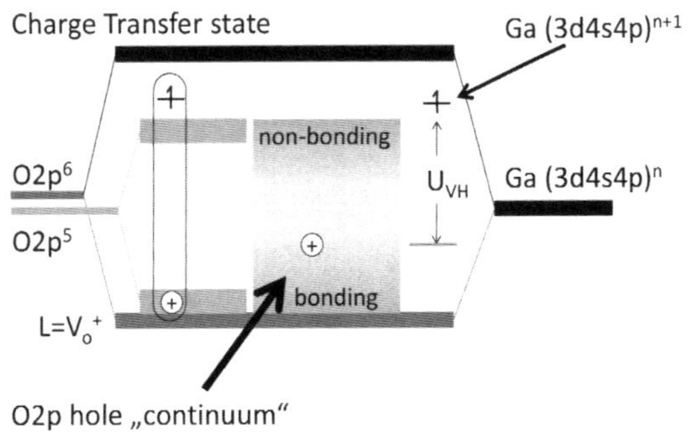

Figure 9: Schematic view of the electronic structure of β-Ga_2O_3 as derived from our spectroscopic analysis. The coexistence of $O2p^6$ and $O2p^5$ states is addressed which forms the O2p hole continuum.

4.3 Model of the electronic structure

At the end of our discussion section we give a schematic explanation of the electronic structure as displayed in figure 9. As discussed above the covalent bands are formed between $O2p^6$ and metal $Ga3d^{10}4sp^0$ states. This interaction causes the electronic band gap. Most important, we have to consider the covalent interaction of $O2p^5$ defect states (small polarons) in addition. This gives the shadowed range in the VB. Within that regime there are the $O2p^5$ states which show a weaker bonding behavior but are distributed all over the VB-CB regime (gap but also including the VB and the CB). These $O2p^5$ defect states form a hole "continuum", hence. Their contribution is smeared out over the complete gap making this material to an n-type semiconductor.

The cigar-like labeled feature in figure 9 describes the CT process where the hole is somewhere in the $O2p^5$ continuum and the electron is right above E_F in the localized Ga4sp states A charge transfer will lift a hole out of the VB into this continuum and simultaneously will pull down the electronic state by the U_{vh} Coulomb attraction. As a consequence we find that the states above the Fermi energy are localized and correlated charge transfer states while the covalent contributions become evident at about 2eV above E_F only. The diagram also explains that the optical gap needs to be smaller than the electronic gap. Indeed we find for the optical gap a value of 4.95eV while the covalent gap is about 7eV.

5. Conclusion

In summary, there is a rather good agreement of our data with the theoretical models of the electronic structure of ß-Ga_2O_3 as manifested by the value of the VBM position, the width of the CB states, and the loss function. However, we obtain some additional information. We find that the degree of covalent bonding contributions is larger as the pDOS of $O2p^6$ derived states in the CB appears to be much larger than predicted by theory. Also we find that the pDOS of the Ga derived states is not a pure Ga4s band but has considerable admixture of Ga4p states.

Focusing on defects we deduce from the observation of a Fano profile that oxygen derived hole ($O2p^5$) states exist in the VB as well as in the band gap regime with a continuous distribution. In addition we find localized defect states which involve CT transitions between $O2p^5$ and Ga4sp states. These states are identified by the appearance of a 3h-Auger decay at the O1s resonance.

Acknowledgements
We like to acknowledge Z. Galazka, K. Irmscher, and R. Fornari (IKZ, Berlin) for providing single crystals as well as C. Janowitz (Humboldt University, Berlin) for discussions. We also appreciate the continuous support of the BESSY staff. This work is supported by BMBF (05KS7IC1, 13N9431).

References

[1] Galazka Z, Uecker R, Irmscher K, Albrecht M, Klimm D, Pietsch M, Brützam M, Bertram R,. Ganschow S and Fornari R 2010 Cryst. Res. Technol. **45** pp. 1229-1236
[2] Lovejoy T C, Yitamben E N, Shamir N, Morales J, Villora E G, Shimamura K, Zheng S, Ohuchi F S and Olmstead M A 2009 *Appl. Phys. Lett.* **94** 081906
[3] Mohamed M, Janowitz C, Unger I, Manzke R, Galazka Z, Uecker R, Fornari R, Weber J R, Varley J B, Van de Walle C G 2010 *Appl. Phys. Lett.* **97** 211903
[4] Janowitz C, Scherer V, Mohamed M, Krapf A, Dwelk H, Manzke R, Galazka Z, Uecker R, Irmscher K, Fornari R, Michling M, Schmeißer D, Weber J R and Van de Walle C G 2011 *New Journal of Physics* **13** 085014
[5] Zhou X T, Heigl F, Ko J Y P, Murphy M W, Zhou J T, Regier T, Blyth R I R and Sham T K 2007 *Phys. Rev.* B **75** 125303
[6] Åhman J, Svensson J and Albertsson G 1996 *J Acta Cryst.* **C52** 1336-1338
[7] Yamaga M, Villora E G, Shimamura K, Ichinose N and Honda M 2003 *Phys. Rev.* B **68** 155207
[8] He H, Orlando R, Blanco M A, Pandey R, Amzallag E, Baraille I and Rérat M 2006 *Phys. Rev.* B **74** 195123
[9] Zhang Y, Yan J, Zhao G and Xie W 2010 *Physica B: Condensed Matter* **405** pp. 3899-3903
[10] Schmeißer D, Hoffmann P and Beuckert G, in: Zschech E, Whelan C and Mikolajick T (Eds.): Materials for Information Technology, Devices, Interconnects and Packaging, Springer, London, 2005, 449-460
[11] Schamm S and Zanch G 2003*Ultramicroscopy* **96** 559-564
[12] Goodenough J B 1960 *Phys. Rev.* **120** 67-83
[13] Orita M, Ohta H, Hirano M and Hosono H 2000 *Appl. Phys. Lett.* **77** 4166
[14] Miroshnichenko A, Flach S and Kivshar Y 2010 *Rev. Mod. Phys.* **82** 2257-2298
[15] Fano U 1961 *Phys. Rev.* **124** 68-74
[16] Schirmer O F 2011 *J. Phys. Condens. Matter* **23** 334218
[17] Richter M, Friedrich D and Schmeißer D *Journal of Electron Spectroscopy and Related Phenomena* , submitted
[18] Schmidt S and Schmeißer D *Solid State Ionic*, submitted

IOP Conf. Series: Materials Science and Engineering **34** (2012) 012003 doi:10.1088/1757-899X/34/1/012003

Growth of Cubic Zinc Sulfide on (001) GaP Substrates

G Haas, S Lautenschlaeger, U Roemer, S Eisermann, A Laufer, M Pinnisch and B K Meyer

I. Physics Institute, Justus-Liebig-University Gießen, Heinrich-Buff-Ring 16, 35392 Gießen, Germany

Gunther.Haas@exp1.physik.uni-giessen.de

Abstract. We report on the successful growth of ZnS epilayers on GaP (001) substrates using a CVD setup. The crystal structure of our layers is cubic zinc blende. The structural aspects and surface morphologies of ZnS films with different thicknesses and growth temperatures have been investigated using X-ray diffraction (XRD) and atomic force microscopy (AFM). Furthermore, low temperature photoluminescence (PL) at 4 K was used to determine optical properties. The appearance of two acceptor related transistions was found to correlate with our analysis of impurity incorporation by secondary ion mass spectrometry (SIMS). The high structural quality and low impurity levels are promising for the further development of the material system, aiming at an application as transparent conductive material.

1. Introduction

There are different classes of materials which basically could serve the needs of transparent conductive materials, such as oxides, nitrides, sulfides or selenides. In this work we focus on zinc sulfide, as it has been shown that ZnS may exhibit p-type [1] and n-type [2] behavior. Yet for zinc sulfide, achieving high n- and p-type conductivity is still problematic, but highly desirable. Therefore, a more intensive study of the basic material properties is needed. So far zinc sulfide has not been a subject to such studies. Especially the (unintentional) incorporation of impurities and the resulting optical and electrical features have so far been documented only in a few publications [3, 4, 5, 6]. The main problem of ZnS is that the cubic and the hexagonal phase have almost the same formation energies. That leads to stacking faults and polytypes, which cause the high electrical resistivity in undoped ZnS [2].

It was our aim to investigate the growth parameters of undoped ZnS for a heteroepitaxial growth approach to find a proper set of growth parameters for further insight into the material system ZnS. That also includes the dependence of structural properties on the layer thickness, as both, crystallinity [7] and surface morphology [8], are influenced by the thickness.

2. Experimental

We used a home built horizontal CVD setup to grow ZnS epilayers. Metallic zinc was used as zinc precursor, while H_2S was the sulfur precursor. The zinc was evaporated thermally and transported towards the reaction zone by a nitrogen flow, where the reaction with H_2S took place and the ZnS epilayer was deposited. We used 8 mbar as chamber pressure and deposition temperatures between 400 °C and 700 °C. As substrates we used, due to the good sticking of ZnS on the surface and the rather small lattice mismatch, GaP substrates (001). The lattice mismatch of GaP and cubic ZnS is only 0.8 %.

Published under licence by IOP Publishing Ltd

From XRD measurements we determined the lattice parameters and the full width at half maximum (FWHM) of the ω-scans by using a Bruker Siemens 5000. The layer thicknesses in this work were determined by several methods, such as optical interference, using a Perkin Elmer Lambda 900, and measurements of the cross section by optical and scanning electron microscopy. The setup for the low temperature photoluminescence measurements consists of a Jobin Yvon THR 1000 monochromator and a Nd:YAG laser (266 nm) with a power of 10 mW. To examine the surface morphology, an AIST-NT scanning probe microscope was used.

3. Results and Discussion

3.1. Crystal quality

A series of samples with a constant H_2S flow of 60 sccm and a deposition time of 5 h was grown. The deposition temperature was varied between 400 °C and 700 °C in steps of 50 °C, respectively 25 °C between 500 °C and 650 °C. To check on the crystalline quality, we performed ω-scans of the (001) reflex. The FWHM of the ω-scan represents the alignment of the ZnS grains along the (001)-axis (tilt), so a lower FWHM value indicates a higher crystallinity along the (001)-axis. Figure 1 shows the full width at half maximum of the ω-scans and the layer thickness plotted as a function of the growth temperature. The best results were achieved at a growth temperature of 550 °C.

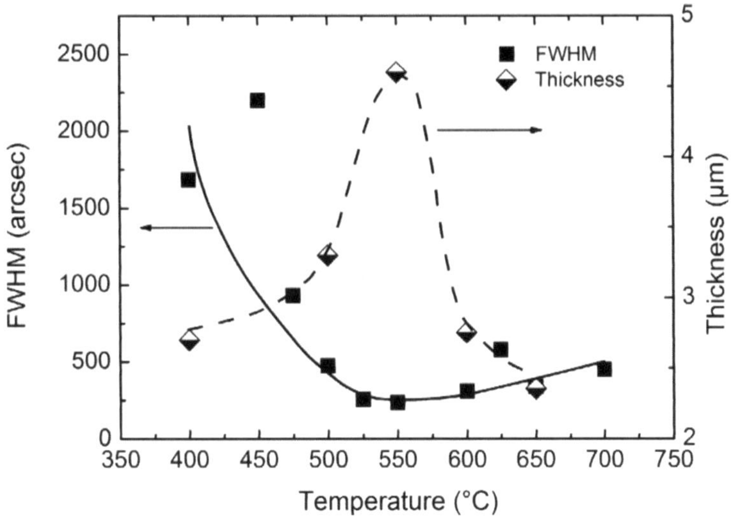

Figure 1. Dependence of the FWHM [ω-scan (001)] and the layer thickness on the growth temperature.

Since it is known that the layer thickness also affects the crystalline quality [7], we investigated if there is a correlation between the layer thickness and the crystallinity. One can see that a growth temperature of 550 °C leads to the thickest films as well as to the best crystalline quality. The FWHM in figure 1 strongly increases for temperatures lower than 550 °C, which leads to an asymmetric curve around the minimum at 550 °C. In contrast to this, the behavior of the layer thicknesses is more symmetric around the maximum at 550 °C. We therefore suggest that not only the growth temperature, but also the layer thickness plays a significant role on the crystal quality. Especially lower growth temperatures (400 – 450°C) give rise to inferior crystal quality.

Figure 2 shows the effect of the layer thickness on the crystalline properties at a growth temperature of 550 °C, figure 2a the FWHM of the ω-scan (001) and figure 2b the lattice parameter of the films. For layers thinner than ca. 250 nm the lattice constant is strongly affected by the substrate, resulting in a coherent lattice parameter and stressed films, similar to the report of Mitsuishi et al. [9]. As the stress starts to release for layers thicker than 250 nm, the lattice constant shifts from coherent to ZnS bulk like. Simultaneously the FWHM starts to rise, indicating a lower crystal quality introduced through the creation of dislocations [10]. Since the dislocations disappear with increasing layer thickness, not only the lattice constant of the layers, but also the FWHM starts to saturate. For a layer of 15.5 μm thickness we achieved a FWHM of 150 arcsec, which is to our knowledge the best value reported for heteroepitaxial ZnS grown on GaP (001) so far.

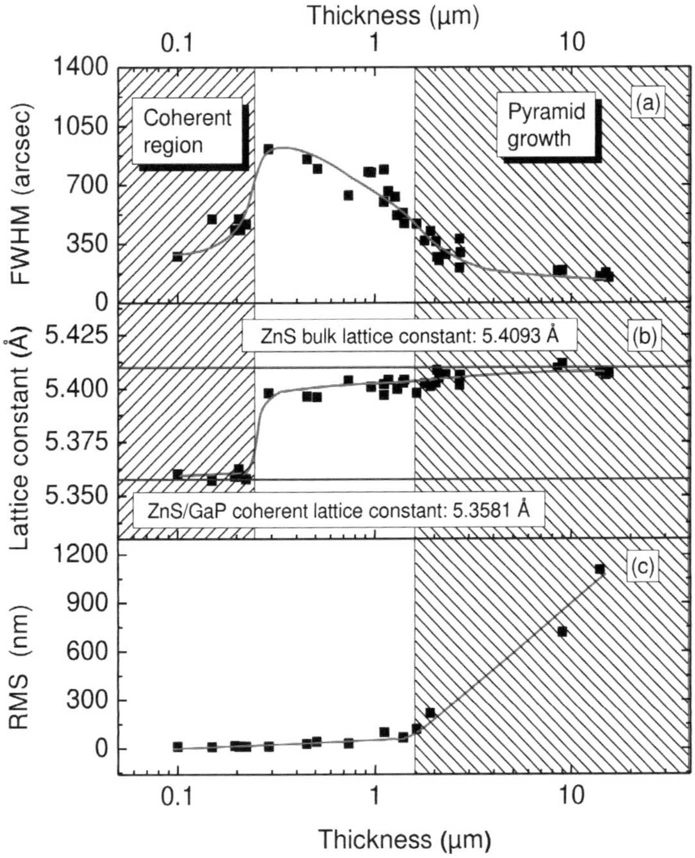

Figure 2.
Layer thickness at a growth temperature of 550 °C versus (a) FWHM of the ω-scan (001), (b) lattice parameter of the films measured by XRD and (c) root mean square roughness of the layer surface according to AFM.

3.2. AFM Measurements

Notable variations in crystal properties, as mentioned above, are often associated with changes in the surface structure, such as variation in crystallite sizes. Regarding this, figure 3a-d gives an insight on the various changes of the surface morphology going along with the layer thickness at a growth temperature of 550 °C. Our investigations using AFM show that, for layers thinner than about 1 μm, the surface consists of small crystallites with diameters of around 200 nm as shown in figure 3a. It is seen in figure 3b that these crystallites start to aggregate once the layer is grown thicker than 1 μm. Figure 3c shows that, for films thicker than 1.5 – 2.0 μm, a complete change in surface morphology has occurred due to the aggregation. Deep pits are starting to appear. These pits turned out to be inverted pyramids, which enlarge with increasing layer thickness (see figure 3d), which agrees well with the findings of Nakamura et al. [11]. Regarding the root mean square (RMS) roughness of the layer surfaces only a small increase for layers thinner than 1 μm occurs, as can be seen in figure 2c.

Once it comes to pyramid like growth, the RMS starts to rise enormously. Thus the decrease of the FWHM for layers thicker than ca. 1.5 μm goes along with the growth of pronounced big pyramid like structures. Figure 3e substantiates that assumption. It shows the surface of a 4.43 μm thick layer grown at 450 °C. Instead of pyramid like features, a rough disturbed morphology results. We conclude that this is the main reason for the low crystalline quality appearing in figure 1 at growth temperatures lower than 500 °C.

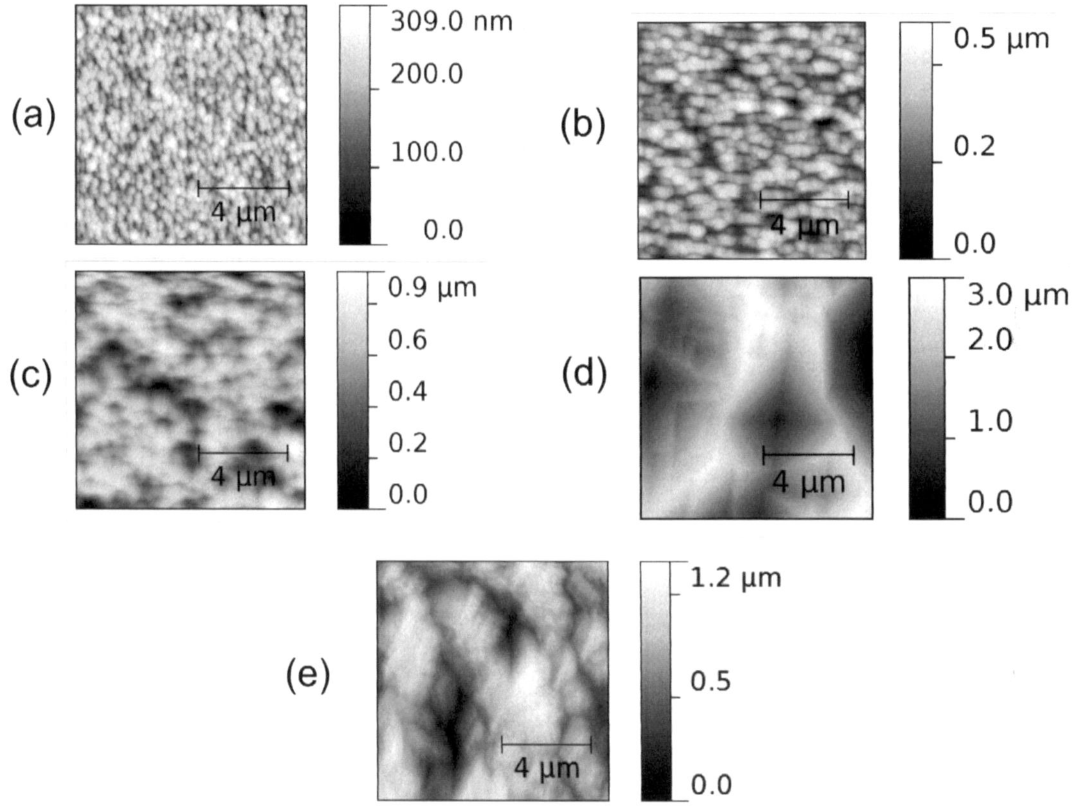

Figure 3. Atomic force microscopy of layers grown at 550 °C with thicknesses of (a) 0.68 μm, (b) 1.38 μm, (c) 1.83 μm, (d) 15.5 μm and (e) a layer grown at 450 °C, 4.43 μm thick.

3.3. Photoluminescence and SIMS

Low temperature photoluminescence measurements were performed to investigate the optical properties of the films. For the interpretation of heteroepitaxial grown ZnS, one has to be aware of the shift in intensity and energetic position that different PL-features undergo, as reported by Nam et al. [10]. In figure 4 we therefore compare layers of similar thicknesses (around 2.5 μm), grown at different temperatures (550 °C, 650 °C). The main difference between both spectra is the disappearance of the donor bound exciton (D^0, X) at 3.785 eV, as the growth temperature is raised to 650 °C. Instead of the (D^0, X), the acceptor bound exciton $(A^0, X)_1$ at 3.780 eV becomes dominant, and another acceptor bound exciton $(A^0, X)_2$ at 3.765 eV appears. Kawakami et al. [5] assigned an acceptor bound exciton at 3.782 eV to Na, while Yodo et al. [2] found a variety of additional PL features introduced through Li+ implantation, including an emission at 3.762 eV. We therefore

checked for the incorporation of these two elements using SIMS (figure 5). The zinc signal is used as a reference. The sputter time correlates to the depth into the layer.
The signals of both elements (Na, Li) rise by an order of magnitude (caused by the higher growth temperature) and therefore possibly introduce the acceptor related features of the PL spectra. There is also a gradient in concentration of the Li- and Na-signal. However, this gradient might be related to surface effects.

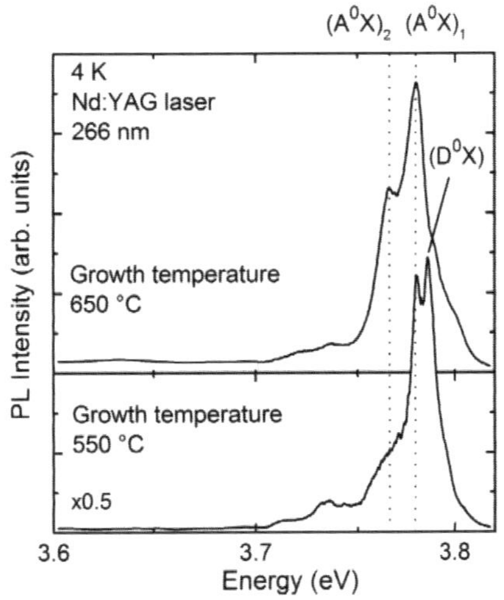

Figure 4. Low temperature PL spectra at 4 K for layers grown at 550 °C and 650 °C.

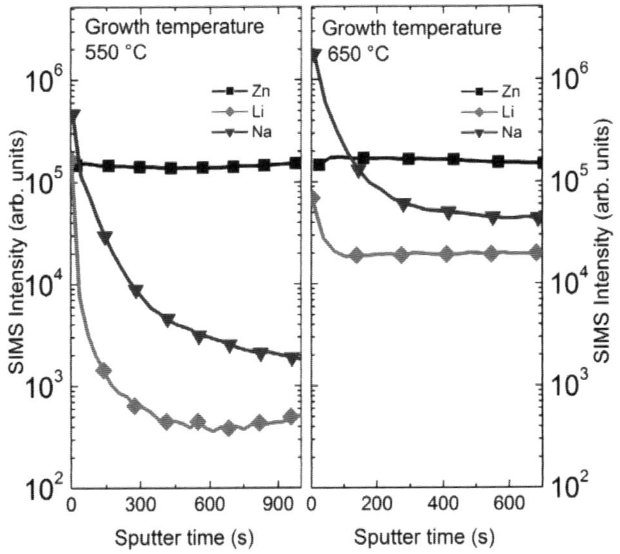

Figure 5. SIMS spectra of layers with different growth temperatures (550 °C, 650 °C). The zinc signal is used as a reference signal.

4. Conclusion

We have shown that the heteroepitaxial growth approach of ZnS on GaP single crystals, using metallic zinc and H_2S as precursors, leads to a successful growth of high quality ZnS epilayers. The influence of the growth conditions, regarding substrate temperature, has been investigated in detail. We have shown that the layer thickness strongly influences the crystalline quality, which is believed to originate from stress and dislocations, caused by the lattice mismatch of the substrate. We have been able to grow layers with a FWHM of the ω-scan as low as 150 arcsec at a growth temperature of 550° C.

The effect of the layer thickness on the surface structure was investigated by AFM. The morphology switches from smaller crystallites to pronounced pyramidal structures for layer thicknesses around 1.5 μm.

In low temperature photoluminescence measurements, acceptor bound excitons are observed correlating with the SIMS measurements, which indicate Na and Li acting as acceptors like impurities.

Despite the formation of the pyramid like structures, (001) seems to be a growth direction of good choice when it comes to crystalline quality and cubic crystal phase. The so far undoped epilayers are a good starting point for upcoming experiments, like bipolar doping.

Acknowledgements
We thank Prof. Dr. Schlettwein, Applied Physics Institute Gießen, for granting us access to his atomic force microscope. This work was supported by the EU project ORAMA FP7-NMP-2009-LARGE-3 NMP-2009-2.2-1, Project Reference No.: 246334.

References

[1] Svob L, Thiandoume C, Lusson A, Bouanani, Marfaing Y and Gorochov O 2000 *Appl. Phys. Lett.* **73** 1695-97
[2] Yoneda K, Toda T, Hishida Y and Niina T 1984 *J. Crystal Growth* **67** 125-34
[3] Yodo T. and Tanaka S. 1992 *J. Crystal Growth* **117** 415-19
[4] Yoneta M, Saito H, Ohishi M, Kitani K, Kobashi H and Hatano C 1995 *J. Crystal Growth* **150** 817-22
[5] Kawakami Y, Taguchi T and Hiraki A 1988 *J. Crystal Growth* **89** 331-38
[6] Taguchi T, Kawazu Z, Ohno T and Sawada A 1990 *J. Crystal Growth* **101** 294-99
[7] Mitsuhashi H, Mitsuishi I, Mizuta M and Kukimoto H 1985 *Jpn. J. Appl. Phys.* **24** L578-80
[8] Vasco E, Zaldo C and Vázquez L 2001 *J. Phys.: Condens. Matter* **13** L663-72
[9] Mitsuishi I, Mitsuhashi H and Kukimoto H 1988 *Jpn. J. Appl. Phys.* **27** L15-17
[10] Nam S, Yu Y-M, O B, Lee K-S, Choi Y D, Jung Y-J 1999 *Appl. Surf. Sci.* **151** 203-12
[11] Nakamura S, Takagimoto S, Ando T, Yamada Y and Taguchi T 2001 *Jpn J. Appl. Phys.* **40** 6993-97

IOP Conf. Series: Materials Science and Engineering **34** (2012) 012004 doi:10.1088/1757-899X/34/1/012004

High mobility annealing of Transparent Conductive Oxides

M Warzecha[1,*], J I Owen[1], M Wimmer[2], F Ruske[2], J Hotovy[1,3] and J Hüpkes[1]

[1]Forschungszentrum Jülich GmbH, IEK5-Photovoltaik, 52425 Jülich GERMANY

[2] Helmholtz-Zentrum Berlin für Materialien und Energie – Institut für Silizium Photovoltaik, 12489 Berlin, GERMANY

[3] Slovak University of Technology, Faculty of Electrical Engineering and Information Technology, Ilkovicova 3, 812 19 Bratislava, SLOVAKIA

E-Mail: [*]m.warzecha@fz-juelich.de

Abstract. To improve electrical properties a high temperature annealing treatment was applied to several transparent conductive oxides (TCO), namely tin doped indium oxide (ITO), Ga- or Al- doped ZnO (ZnO:Al/Ga), ion beam assisted deposited (IBAD) ZnO:Ga and Ga doped zinc magnesium oxide (ZnMgO:Ga). All these films were grown by magnetron sputtering. During the annealing process all TCO films were capped with 50 nm of amorphous silicon in order to protect the films from environmental impact. Increase in mobility up to 72 cm^2/Vs and low resistivity of 1.6×10^{-4} Ωcm was achieved for ZnO:Al after annealing at 650°C for 24 h. Independent of the deposition conditions and doping or alloying material almost all ZnO based films show a consistent improvement in mobility. Also for ITO films a decrease in resistivity with partially improved mobility was found after annealing. However, not all ITO films show consistent improvement, but carrier density above 10^{21} cm^{-3} while ZnO films show no clear trend for carrier density but a remarkable increase in mobility. Thus we propose the healing of defects and the activation of donors to be most significant effects for ZnO and ITO films, respectively.

1. Introduction

Highly transparent conductive oxides (TCO) are promising materials especially as window layers for thin film photovoltaic devices. Due to its wide band gap of 3.3 eV and a resistivity as low as 2×10^{-4} Ωcm after doping, ZnO:Al is the most favored window material for thin film silicon solar cells [1,2]. Magnetron sputtered ZnO:Al layers attracted great attention as transparent electrode for thin-film silicon solar cells due to its low cost, high abundance and nontoxicity.

Excellent conductivity and transparency properties are reported for tin doped indium oxide [3,4]. In addition, ITO has a higher band gap of 3.7 eV and is therefore transparent to a wider spectrum of the solar radiation compared to ZnO. Since indium, the principal material in ITO is scarce and expensive, alternatives have to be found.

Recently, Wei *et al.* [5] reported increase in band gap up to 4 eV by alloying of ZnO with Mg. This makes $Zn_{1-x}Mg_xO$ a possible candidate as a window layer for thin film solar cells since the transparency in the short wavelength range of the solar radiation can be enhanced. However, for high Mg concentrations a phase separation from hexagonal to the cubic structure occurs. Nevertheless, a variation of the band gap between 3.3 eV and 4 eV without phase separation was realized by pulsed laser deposition of $Zn_{1-x}Mg_xO$ thin films [5]. Doped $Zn_{1-x}Mg_xO$ films, grown by magnetron sputtering

Published under licence by IOP Publishing Ltd

revealed high mobility values. However, the mobility and carrier concentration decreased with increasing Mg fraction [6].

Ion beam assisted deposition (IBAD) at room temperature, i.e. growth from a ZnO target with an auxiliary ion gun, leads to improvement in crystal quality which is usually achieved after growth at high substrate temperatures [11].

High mobility annealing of ZnO:Al films at 650 °C under an amorphous Si capping layer leads to an improvement in electrical properties, as recently reported by Ruske *et al.* [13]. Therefore, our work focuses on the investigation of this annealing treatment of several TCO materials. Exploration on the annealing effect of alternative doping and alloying materials for ZnO and ITO as well as improvement in electrical properties are the important tasks for our work.

2. Experimental

ZnO:Al, ZnO:Ga, IBAD-ZnO:Ga, $Zn_{1-x}Mg_xO$:Ga and ITO thin films were prepared by magnetron sputtering in different sputtering setups: An in-line sputtering system VISS 300 *(VAAT, Dresden)* was used to grow ZnO:Al films [7,8].

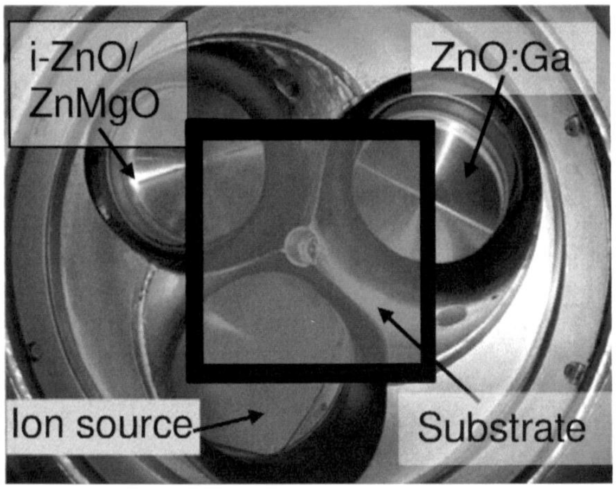

Figure 1. Top view on the sputtering chamber of the cluster tool system for the growth of ZnO:Ga, ZnMgO:Ga and IBAD-ZnO:Ga films. The ion source is covered by a shutter.

ZnO:Ga and $Zn_{1-x}Mg_xO$:Ga thin films were grown in a multi-cathode sputter chamber of a cluster tool system *(VAAT, Dresden)*. Figure 1 shows the top view of this sputtering chamber revealing three cathodes inside which can be operated independently for simultaneous sputtering or ion beam assisted deposition (IBAD). In the center of the three targets a schematic picture of the substrate is indicated. During deposition the substrate rotates with a rotation velocity of 5 r/min in order to achieve homogeneous layers. The variation in the Ga content was realized by operating two cathodes simultaneously with a ZnO:Ga_2O_3 ceramic target (98:2 w/w%) and an intrinsic ZnO target from *Kurt J. Lesker* company, operated in DC and RF (13.56 MHz) mode, respectively. Changing the target power led to change in the deposition rate and thereon the Ga concentration in the films could be varied between 0 and 2 wt. %.

ZnMgO:Ga films were grown in the same way using the ZnO:Ga_2O_3 and an ZnO:MgO ceramic target (50:50 w/w%). During deposition the pressure was set to 3.3 x 10^{-3} mbar and the substrate temperature to T_S = 340 °C, except for one $Zn_{0.89}Mg_{0.11}O$:Ga sample which was grown at T_S = 250 °C.

The Mg and Ga contents of the $Zn_{1-x}Mg_xO$:Ga films were determined by the deposition rate of the targets. Verification of these values has been done by wavelength-dispersive x-ray microanalysis spectrometry (WDX) in a *Cameca SX100* system.

Since annealing even in vacuum at high temperatures generally leads to a deterioration of the electrical properties [9], an amorphous silicon capping layer of about 50 nm thickness was deposited by PECVD on top of the TCO films prior to annealing in order to protect the films from the detrimental influence of the environment [13]. For the annealing process the films were put in a furnace which was afterwards evacuated to a base pressure of 3 x 10^{-6} mbar. The heating temperature was set to 650 °C for 24 hours. The heating rate was 21 K/min for the first annealing studies but was later reduced to 2.6 K/min. After heat treatment the films were left for 8 hours in the furnace, i.e. the cooling rate was 1.3 K/min.

For the growth of ion beam assisted (IBAD) ZnO:Ga thin films the ZnO:Ga (2 wt. %) target and an auxiliary ion source were put into operation [10,11]. The ion source in our sputter chamber is an inverse sputter etching tool ISE 90 *(VAAT, Dresden)*. Due to the coeval operation of a ZnO target and an ion source a simultaneous growth and bombardment of the film surface is achieved to provide additional momentum transfer to the growing film. This should replace or support external substrate heating. ZnO:Ga thin films with varying ion source power between 40 and 150 W were grown by IBAD. During the deposition the pressure was set to 3.3 x 10^{-3} mbar for all IBAD films.

ITO thin films were grown in a static deposition system by *Kurt J. Lesker* Company where the ceramic ITO target was arranged directly below the substrate. A random distribution of ITO films with different growth conditions such as different substrate temperatures, target powers, O_2 flow ratios, film thicknesses as well as deposition pressures were selected to study the annealing effect on these films.

Hall measurements before and after annealing were collected using the van der Pauw method in a *Keithley 926* Hall setup.

In order not to distort subsequent Hall measurements capping layers of most of the films were removed in an NF_3 plasma dry etching process. Some films were measured with a capping on top, but no effect on the electrical properties was observed.

Transmission spectra of several Ga doped IBAD-ZnO and $Zn_{1-x}Mg_xO$ films before and after annealing were collected with a dual beam spectrometer (*Perkin Elmer*, Lambda 19) with a wavelength range between 250 nm to 2000 nm. To measure the transmittance, capping layers of these films were finally removed by the NF_3 plasma dry etching process.

Determination of the band gap energies E_g of $Zn_{1-x}Mg_xO$:Ga films was done by plotting $(\alpha h\nu)^2$ vs. $h\nu$ according to the relationship $\alpha^2 \sim (h\nu - E_g)$, where $h\nu$ is the photon energy and α the absorption coefficient, respectively [12].

X-ray diffraction measurements were taken in a Θ-2Θ geometry utilizing a *Phillips X'Pert Pro* diffractometer with CuK_α radiation (λ = 0.154 nm).

3. Results and discussion

3.1. Transfer of cap preparation and annealing treatment

High mobility annealing studies have already been done by Ruske *et al.* [13] at the Helmholtz Zentrum Berlin. These results are briefly summarized by the red points in Figure 2 which shows (a) the mobility, (b) carrier concentration and (c) resistivity of Al doped ZnO thin films after 24 h of annealing at temperatures between 500 and 650 °C. Black points in Figure 2 are measured data obtained after annealing similar films utilizing a heater setup and cap deposition in Jülich. Examining the experimental results obtained by Ruske *et al.* in Figure 2(a), it is seen that annealing leads to a significant increase in mobility from 42 cm^2/Vs for the as deposited sample up to 67 cm^2/Vs for the sample which was annealed at the highest temperature of 650 °C. Comparing this result with first annealing measurements in Jülich the same annealing effect is observed with the mobility increasing from 49 cm^2/Vs to 72 cm^2/Vs after annealing at 650 °C. The higher mobility observed after annealing in Jülich compared to the measurements by Ruske *et al.* are likely related to slight differences in the measurement setup.

Carrier concentration (Figure 2(b)) increases with annealing from 5 x 10^{20} cm^{-3} to about 6 x 10^{20} cm^{-3} upon annealing. The new series, however, shows a decrease down to about the initial value for annealing at temperatures above 500°C. This might indicate some limitation of the barrier properties of the cap, but changes in carrier density are quite small compared to the mobility values and variations observed upon annealing without cap [9].

The resistivities of both annealing series (Figure 2(c)) lie very close together. A decrease in resistivity from about 2.5 x 10^{-4} Ωcm for the reference sample down to 1.5 x 10^{-4} Ωcm at an annealing temperature of 600 °C is obtained for our films. At 650 °C a deviation from the literature data is seen: While the resistivity for the films of Ruske *et al.* decreases down to 1.4 x 10^{-4} Ωcm, the films annealed in Jülich show an increase in resistivity up to 1.8 x 10^{-4} Ωcm which is still below the initial value.

All in all, this annealing procedure under a protection layer leads to a raise in mobility for our high quality ZnO:Al films and the procedure is easy to transfer to other labs. Deviations in the measurements of carrier concentration and resistivity can be possibly attributed to a non-optimized heating process in our institute, such as a too short heating ramp. In this case the films and the capping layer are exposed to a very high temperature after a short period of time, in our case it is 30 min. During this short period of time an out diffusion of the H atoms is not likely, instead an accumulation at voids and dislocations can lead to a local peeling off of the capping layer. Finally, due to this effect the TCO film is not protected from the environment anymore and degradation of the electrical properties can be explained by this effect.

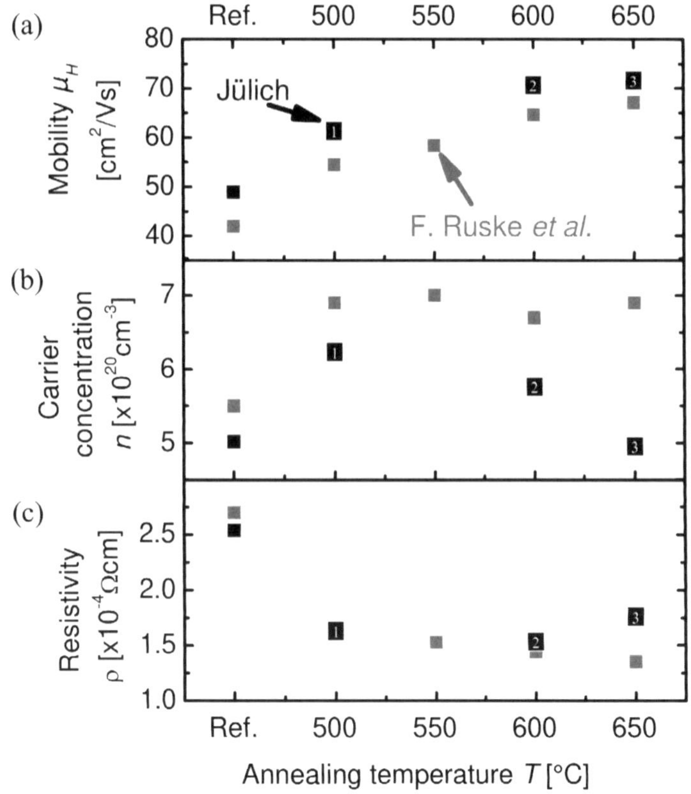

Figure 2. Transfer of the high mobility annealing procedure. Annealing temperature dependencies: (a) mobility, (b) carrier concentration and (c) resistivity of Al doped ZnO thin films after 24h of annealing for different temperatures. The red measurement points are taken from Ruske *et al.* [13] and the black squares show the transfer to Jülich. Numbers inside the measurement points will be reused in Figure 4.

In this regard, the heating rate was reduced for the following annealing experiments down to 2.6 K/min as mentioned in section 2. The heating time was still kept at 24 hours.

3.2. Effect of ZnO:Al film thickness on high mobility annealing

The dependence of high mobility annealing on a film thickness series and hence on the grain size of ZnO:Al thin films is investigated in this section.

Figure 3 shows (a) the resistivity, (b) the mobility and (c) the carrier concentration as a function of film thickness before (black full squares) and after (red open squares) annealing. As the film thickness increases, the mobility in Figure 3(a) shows a drastically increase from 13 cm^2/Vs to 43 cm^2/Vs at a film thickness of about 400 nm. Increasing the film thickness further yields only to a relatively small increase in mobility up to 50 cm^2/Vs. Therefore, saturation in mobility for film thicknesses higher than 400 nm is observed which is denoted by an asymptotic line in the graph.

The same dependence is seen for the carrier concentration (Figure 3(b)). Increasing the film thickness leads to an increase in carrier concentration from 3 x 10^{20} cm^{-3} to 5 x 10^{20} cm^{-3} up to a film thickness of 400 nm. Further increasing the film thickness shows only comparatively small variation in carrier concentration from 4 x 10^{20} cm^{-3} to 5 x 10^{20} cm^{-3} at a film thickness of 1575 nm. Similar trends have been observed by others [14].

Since mobility and carrier concentration increase, a decrease in resistivity with increasing film thickness is seen in Figure 3 (c). Similar to mobility saturation in resistivity at 400 nm with a value of about 2.5 x 10^{-4} Ωcm is shown.

After annealing the saturation level shifts to higher values for the mobility and to smaller values for the resistivity. Therefore, highest mobility of 72 cm^2/Vs and lowest resistivity of 1.6 x 10^{-4} Ωcm are obtained after annealing.

The carrier concentration, on the other hand, does not show any clear trend upon annealing: Some films show a decrease and some an increase. A change of less than 20 % is obtained due to annealing and no correlation to the film thickness is observed. This behavior is not clear and is still under investigation.

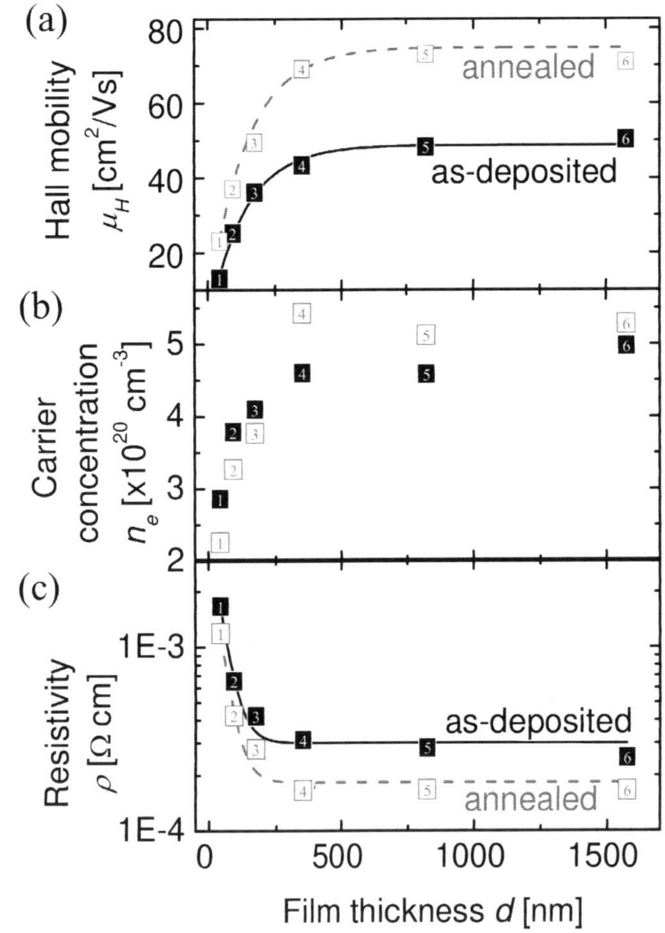

Figure 3. Electrical properties of Al doped ZnO thin films before (black squares) and after (open red squares) high mobility annealing: (a) Mobility, (b) carrier concentration and (c) resistivity as a function of film thickness. The solid and dashed lines are meant to guide the eye. Numbers inside the measurement points will be reused in Figure 4.

3.3. Annealing effect in ZnO:Al

For a better understanding of the electrical properties of TCO films a collection of the mobility data for different polycrystalline ZnO:Al films, grown under various sputtering conditions is summarized in Figure 4 where the mobility is plotted against the carrier concentration before (full symbols) and after annealing (open symbols).

Blue measurement points are taken from the first annealing studies of ZnO:Al as shown in Figure 2. Therefore, only one blue full measurement point is shown as the reference sample and the three open squares belong to the annealed films. Black squares with numbers inside show the film thickness series of ZnO:Al taken from Figure 2.

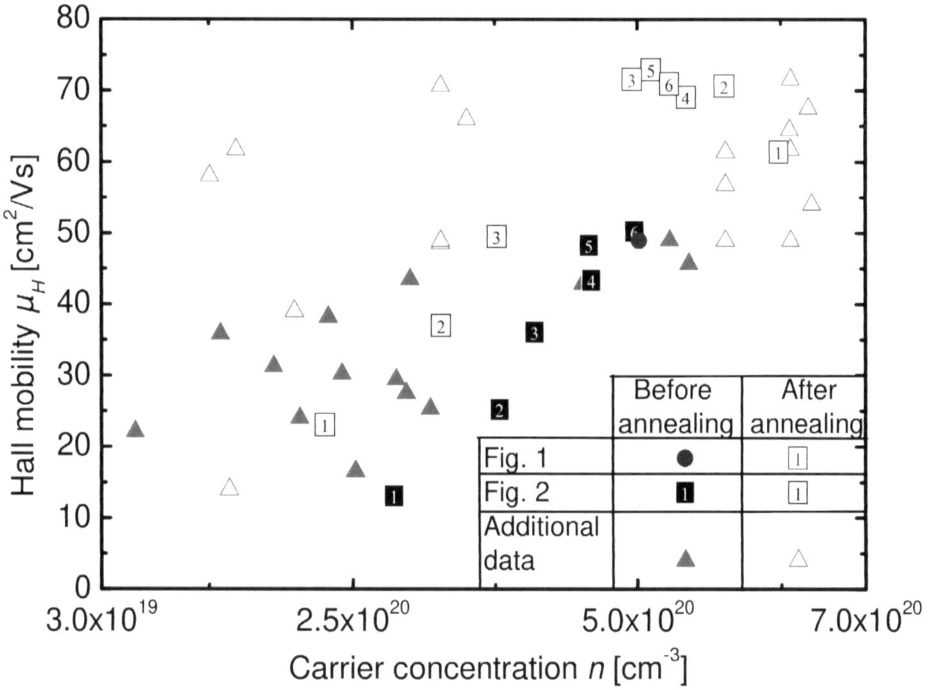

Figure 4. Hall mobility data as a function of carrier concentration for Al doped ZnO thin films grown at various sputtering conditions. The measurement points are indicated with a number to allow the reader find the sample after annealing. The blue solid circle and the squares with numbers inside are taken from Figure 2, black squares from Figure 3 and red triangles represent additional ZnO:Al films prepared under various conditions with different doping levels [8].

The red triangles in this graph indicate additional high mobility data which was measured before (red full triangles) and after annealing (red open triangles) [8].

For small film thicknesses the films mainly consist of a nucleation zone where small grains with varying crystal orientations dominate the polycrystalline film [14]. Due to the survival-of-the-fastest growth model [15] grains with high growth rates, which are those who grow in the c-axis direction, overgrow others. This leads to improved texture and alignment of c-axis along substrate normal. After the nucleation zone, the grain size cannot be increased further and the mobility seems to be limited by the limited grain size. Thus we consider grain boundary scattering and scattering at structural defects as the major scattering mechanism.

Increasing the mobility far above the previous experimental limit upon annealing hints to reduction of defect density as major effect. A change in grain size can be excluded since no recrystallization during the annealing process is observed [16].

Generally, high mobility annealing leads to increase in mobility and likewise slight increase in carrier concentration for ZnO:Al is seen.

One sample in Figure 4 of the additional ZnO:Al series show degradation upon annealing since the mobility value of 14 cm^2/Vs of the annealed sample is quite below the data of the residual as-deposited ZnO:Al. Crack formation during annealing could be a possible explanation for this deterioration.

3.4. Annealing of ZnO based alloys

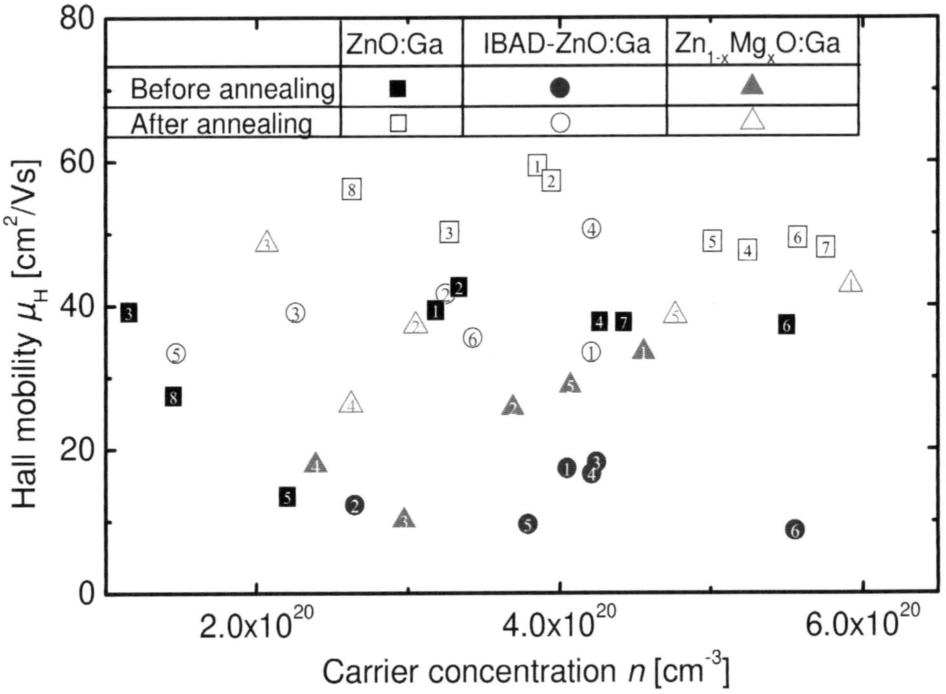

Figure 5. Hall mobility as a function of carrier concentration for various ZnO:Ga, IBAD grown ZnO:Ga and $Zn_{1-x}Mg_xO$:Ga thin films before and after annealing. The numbers inside the measurement points should allow the reader to recover the samples before and after annealing. These numbers also serve as identification numbers for each sample. More information about the deposition conditions of and the Mg contents in these films can be found in Table 1.

To get further insight into the annealing effect we investigated the influence of the alternative dopant gallium for ZnO as well as Ga doped $Zn_{1-x}Mg_xO$ as an alloy prepared under various conditions.

Figure 5 shows the Hall mobility and carrier concentration dependencies for ZnO and $Zn_{1-x}Mg_xO$ thin films before and after annealing. Black squares show the mobility of films grown at various sputtering conditions while the blue circles belong to a series of ion beam assisted deposition (IBAD) of ZnO:Ga [10]. Annealing dependencies of $Zn_{1-x}Mg_xO$:Ga thin films are shown by red triangles.

For completeness, each measurement point of the three series is marked with a number to assign each sample before and after annealing to the specific deposition parameters. These numbers represent also sample identification numbers. The corresponding deposition conditions and material properties are given in Table 1 and figures 2 – 4, respectively.

The first column in this table shows the sputtering conditions for the ZnO:Ga films. ZnO:Ga films were grown at various sputtering conditions and with different Ga contents and annealed in order to determine the influence Ga doping and sputtering conditions.

The full numbered data point represents the same sample after its subsequent annealing process. Similar to Figure 4 for the ZnO:Al thin films all the mobility values before annealing seem to be limited by grain boundary scattering at other structural defects [17]. The experimental results of high temperature deposited films lie in the range between 28 and 43 cm^2/Vs, except sample number 5 which was grown at room temperature and shows the smallest mobility of 14 cm^2/Vs. On the other hand, this film shows the highest increase in mobility from 49 cm^2/Vs upon annealing, meaning an increase by a factor of 3.6 and well above the best as-deposited values. Additionally, the highest mobility after annealing showed sample number 2, starting with an initial value of 39 before and 59 cm^2/Vs after annealing, respectively. In general, ZnO:Ga thin films show the same behavior as ZnO:Al after annealing. This is a hint that the annealing process leads to an increase in mobility independent of the doping material and is at least valid for several different growth conditions. For deposition parameter dependence, no clear trend was observed.

Table 1. Deposition parameters for the three film series shown in Figure 5. ***1st row:*** Ga content, temperature T_S and pressure during the growth of as-deposited ZnO:Ga films. ***2nd row:*** Ratio between the ion source and the ZnO:Ga target power as well as T_S during the growth of IBAD-ZnO:Ga thin films from Figure 5. "Seed layer" means that only 30 nm of ion beam assistance was done, switching off the ion source after 30 nm. ***3rd row:*** Mg concentrations after and substrate temperatures during growth of $Zn_{1-x}Mg_xO$:Ga thin films, respectively.

Sample number from Figure 5	ZnO:Ga			IBAD-ZnO:Ga		$Zn_{1-x}Mg_xO$:Ga	
	Ga content [wt. %]	T_S [°C]	Deposition pressure [mbar]	Power ratio	T_S [°C]	Mg content x	T_S [°C]
1	1	350	1.0	0	RT	0.03	340
2	1	350	20	0.4	RT	0.05	340
3	1	550	3.3	0.5	250	0.08	340
4	1.5	340	3.3	0.5 (seed layer)	RT	0.11	340
5	2	30	3.0	1.5 (1060 nm)	RT	0.11	250
6	2	250	3.0	1.5 (700 nm)	RT		
7	2	340	3.0				
8	2	500	3.0				

IBAD-process has proven to promote c-texture growth of ZnO, but induces severe stress [10,11]. All IBAD-ZnO:Ga films were grown at room temperature except sample 3 which had a substrate temperature of T_S = 250 °C during deposition. Samples 5 and 6 were grown at the same sputtering conditions, however they show different film thicknesses d after growth: While sample 5 reveals a film thickness of 1060 nm, sample 6 is 700 nm thick.

Generally, the IBAD-ZnO:Ga films show comparatively low mobility values of less than 20 cm^2/Vs in the as-deposited sate. This might be caused by the defect formation at low temperature sputtering, but even at high temperature (sample 3, blue) the mobility is lower than the respective reference sample without ion beam assistance (sample 6, black). Thus, low mobility is related to the structural defects induced by the ion bombardment. However, the annealing process leads to an increase of all the values of the IBAD-ZnO:Ga thin films reaching the mobility values of as-deposited ZnO:Ga thin films (black squares) prepared at high deposition temperature.

The IBAD-ZnO:Ga sample 4 in Figure 5 was grown with only 30 nm of an IBAD seed layer. This means that after deposition of 30 nm the ion source was switched off and the ZnO:Ga target operated solely. As can be seen in Figure 5 this sample shows the highest mobility after annealing compared to the residual IBAD-ZnO:Ga films. A mobility of 51 cm^2/Vs is comparable to mobilities for samples which were grown at higher deposition temperatures and additionaly underwent the high mobility annealing process.

To investigate the additional ion bombardment on the crystal structure of the IBAD-ZnO:Ga films, Figure 6 shows the X-Ray diffraction pattern of the (002) reflection peak of the ZnO-lattice for unannealed IBAD-ZnO:Ga films. The black curve shows the diffraction pattern of an ZnO:Ga thin film, grown at room temperature which serves as the reference signal. Increase in the ion gun power up to a factor of 0.4 leads to a remarkable increase in the intensity of the diffraction peak. Further increase in the ion source power to 1.5 diminishes it to a value which is below the maximum of the reference ZnO:Ga film, sputtered without ion beam. This shows that additional ion bombardment at such high energies leads to damage in the crystal structure. In addition to a change in intensity a shift of the diffraction peak to lower diffraction angles with increasing the power ratio is observed. This shift is caused by stress formation in the films due to ion impact. The sample which was grown by an ion beam power ratio of 1.5 reveals also peak splitting, which indicates phase separation into strained and relaxed ZnO phases.

On the other hand, the sample which was grown with only 30 nm IBAD-seed layer reveals almost no change in the intensity and peak position. The sample, grown at a higher substrate temperature of 250 °C and a power ratio of 0.5 leads to a diffraction intensity which is more than one order of magnitude higher than the intensity of the reference sample.

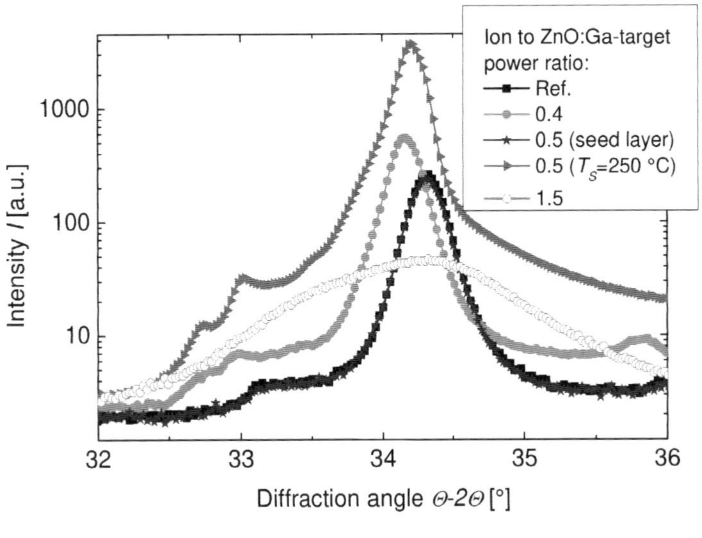

Figure 6. X-Ray diffraction pattern of IBAD grown ZnO:Ga films at different power ratios.

Since ion beam assisted deposition of ZnO:Ga leads to improvement in crystal quality but deterioration in electrical properties due to stress formation [10], we now study the influence of texture parameters on annealing behavior.

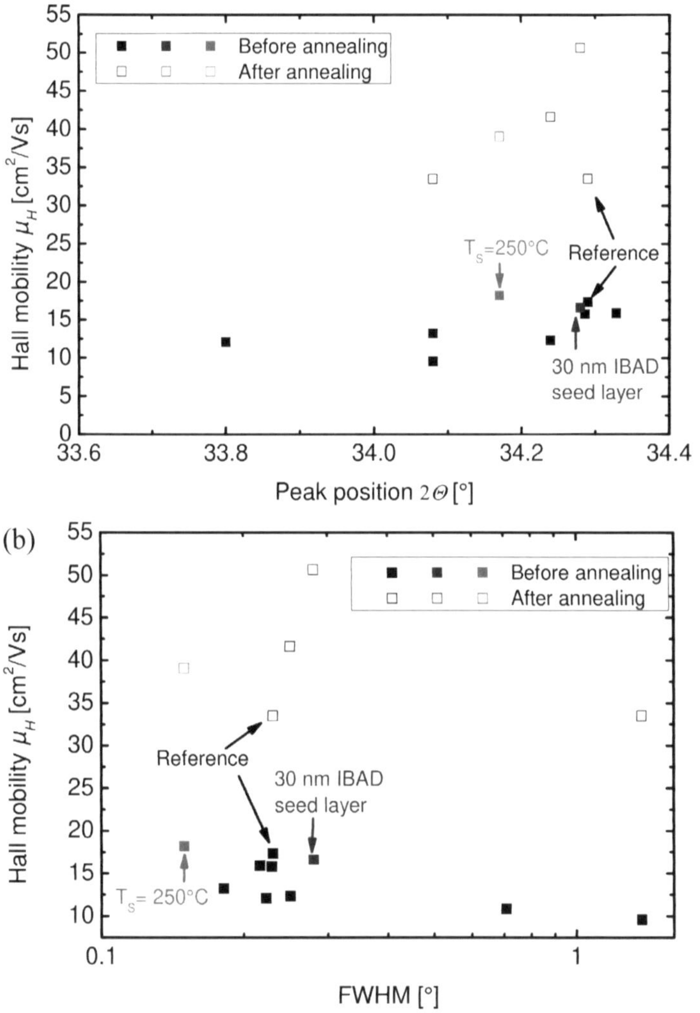

Figure 7. (a) Hall mobility as a function of peak position and (b) as a function of FWHM of the (002) diffraction peak.

Figure 7 (a) shows the mobility in dependence to the XRD (002) diffraction peak position as a measure of film stress for several IBAD grown ZnO:Ga thin films before and after annealing. Sample No. 6 was not analyzed by XRD, hence only five post deposition annealed samples are shown in this graph. Some films which were not annealed but analyzed by XRD are also added to this graph in order to determine the stress effect on IBAD-grown ZnO:Ga films. Before annealing, the reference sample, grown without ion beam assistance, shows the highest mobility of about 17 cm^2/Vs. The peak shift leads generally to a decrease or rather to saturation in mobility below 16 cm^2/Vs, except for the high temperature (T_S = 250 °C) deposited sample with a mobility of 18 cm^2/Vs. Stressed with (002) peaks below the reference sample exhibit lower mobility than the reference sample. However, the highly stressed IBAD-ZnO:Ga films do not follow the trend further. This might be related to phase separation into relaxed phase. The high temperature deposition leads to much higher mobility values as compared to the room temperature sputtered IBAD-ZnO:Ga samples. Annealing increases the mobility for all samples and stress does not show a direct correlation.

The sample consisting of 30 nm IBAD seed layer reveals the highest mobility after annealing. Köhl *et al.* [11] recently have shown that ion bombardment during deposition of ZnO by an auxiliary ion gun only during the initial growth stage leads to an improvement in crystal structure, but do not suffer from increased stress due to peening effect by the additional bombardment. Even though, the seed layer ZnO does not show any difference in XRD or electrical properties before annealing, compared to the reference sample. This might be an explanation for the highest mobility after annealing for IBAD-ZnO film No. 4.

Since the FWHM is typically related to the grain size [18], Figure 7 (b) shows the Hall mobility vs. FWHM of the (002) diffraction reflection before and after annealing. Please note that the FWHM might be limited by structural defects and microstrain, and does not provide information on lateral grain size which is important for electrical transport. Before annealing a slight decrease in mobility with increasing FWHM is seen which can be explained by the worse crystalline quality. After annealing no clear dependence to the FWHM of the as deposited state is noticed.

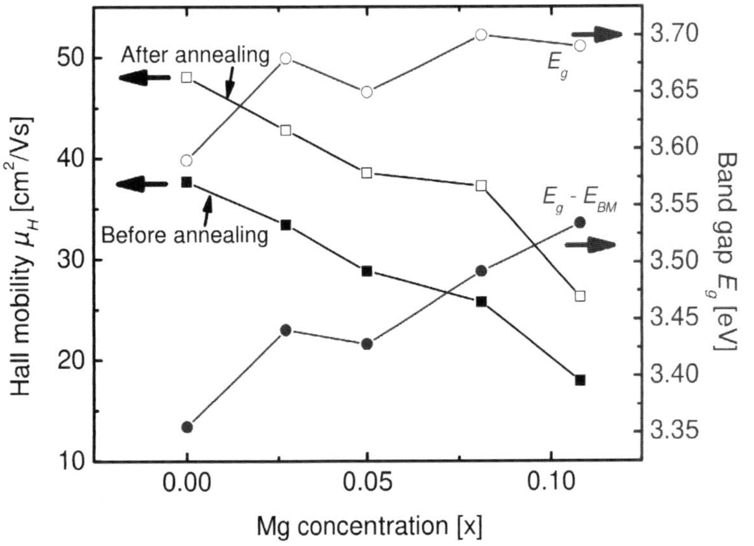

Figure 8. Hall mobility before and after annealing as well as the change in the band gap E_g and the Burstein-Moss corrected band gap E_g-E_{BM}, as a function of Mg concentration for the $Zn_{1-x}Mg_xO$:Ga thin films.

The red triangles in Figure 5 show annealing experiments of $Zn_{1-x}Mg_xO$:Ga thin films in order to determine the effect of Mg incorporation into ZnO:Ga and the subsequent annealing behavior. The mobility values of the as deposited $Zn_{1-x}Mg_xO$:Ga films decrease with increasing Mg content down to less than 20 cm^2/Vs at a Mg content of about 11 %. Annealing procedure leads to an increase in mobility for all $Zn_{1-x}Mg_xO$:Ga films upon annealing and the highest mobility of 49 cm^2/Vs is achieved for the sample with the number 3. Likewise, this sample shows also the highest change in mobility from 10 cm^2/Vs to 49 cm^2/Vs, which means a change by a factor of almost 5.

Figure 8 shows the dependence of the Hall mobility on the Mg content in the $Zn_{1-x}Mg_xO$:Ga films before and after annealing. The blue line shows the change in band gap with increasing Mg content. Sample No. 5 of the $Zn_{1-x}Mg_xO$:Ga series was excluded in this graph because T_S during deposition differs. Since the Mg content leads to lower carrier density and Burstein-Moss shift (E_{BM}) [19] compensates the increased band gap we show the electronic (E_g) and optical (E_g-E_{BM}) band gaps. An increase in the electronic band gap from 3.36 eV to 3.48 eV for $Zn_{0.89}Mg_{0.11}O$:Ga is observed. Annealing leads to a shift of the mobility towards higher values but the slope sustains. Assuming that no change in band gap upon annealing occurs, improvement in electrical properties due to annealing and a wider band gap a towards the UV due to incorporation of Mg into ZnO can be achieved.

In conclusion, nearly all ZnO based films, independent on the doping material and alloying show a significant increase in mobility upon annealing even up to a factor of 4. Opposite to thick ZnO:Al and ZnO:Ga films, decrease in carrier concentration was observed for some $Zn_{1-x}Mg_xO$:Ga films as well as for some IBAD-ZnO:Ga films. Sample 2 and 3 of the $Zn_{1-x}Mg_xO$:Ga in series show decrease in carrier concentration. However, there is no correlation to the Mg content as can be read from Table 1. The same is true for severely stressed IBAD-ZnO:Ga series. Since the additional ion bombardment induces stress in the films, this anomalous behavior in carrier concentration might be attributed to stress effects in these films. Stress leads to an increase in defect and dislocation density. A suggestion for a decrease in carrier concentration could be assigned to the healing off defects and dislocations upon annealing.

3.4.1. Optical properties

The effect of high mobility annealing on the transmission spectra for IBAD grown ZnO:Ga (samples 1-4 from Table 1) is shown in Figure 9. In the as-deposited state, some of these films show quite poor transmission. After annealing, all the films show improvement in transmittance almost for the whole spectral range. The sample, grown at T_S = 250 °C (Figure 9, bottom) shows high transmittance up to 80 % in the visible spectral range already after deposition. However, the effect of annealing is the increase in transmittance in the short and long wavelength range of the spectrum by a much steeper absorption edge and shift of free carrier absorption. This is a hint that the defect density, which causes energy levels in the bandgap, is strongly reduced upon annealing.

It is interesting to note that the seed layer sample (Figure 9, top, solid lines) almost shows the same transmission spectra before and after annealing as the reference ZnO:Ga sample without ion beam assistance (Figure 9, top, dashed lines), leading to the conclusion that the seed layer has no significant influence on the optical properties.

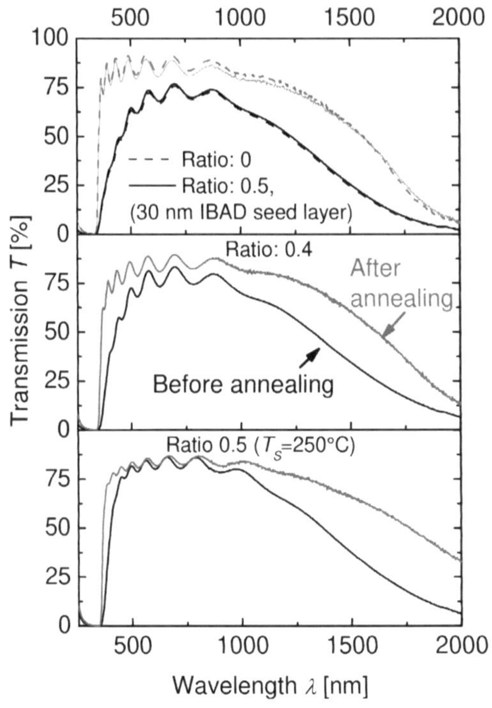

Figure 9. Transmission spectra of IBAD-ZnO:Ga films grown at different power ratios before and after annealing.

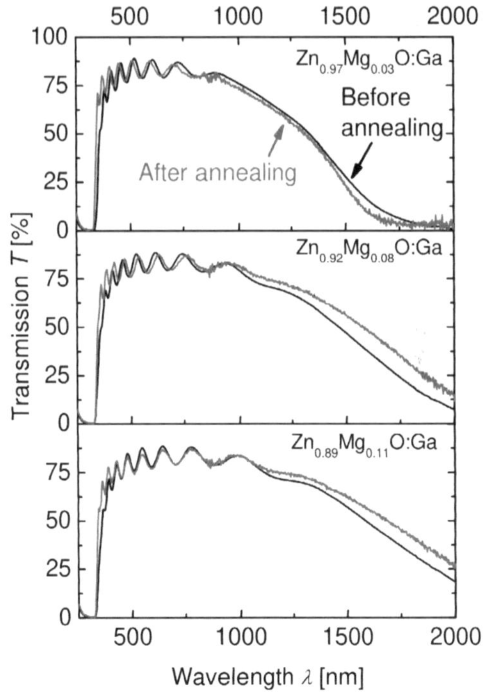

Figure 10. Transmission spectra of $Zn_{1-x}Mg_xO$ thin films with varying Mg contents before and after annealing.

The annealing effect on $Zn_{1-x}Mg_xO$:Ga films (samples 1, 3 and 4 from Table 1) is indicated in Figure 10. In correlation to the IBAD-ZnO:Ga films an increase of the transparency in the short wavelength spectral range upon annealing is achieved for all these films, making the absorption edge steeper. However, this change is not as significant as the increase of the transparency for the room temperature grown IBAD-ZnO:Ga films.

IOP Conf. Series: Materials Science and Engineering **34** (2012) 012004 doi:10.1088/1757-899X/34/1/012004

Changes of the optical properties in the long wavelength range (> 1000 nm) can usually be explained by changes in carrier concentration upon annealing. Another explanation could be an improved mobility which will reduce the impact of free carrier absorption to the visible and NIR spectral range.

3.5. Annealing of ITO films

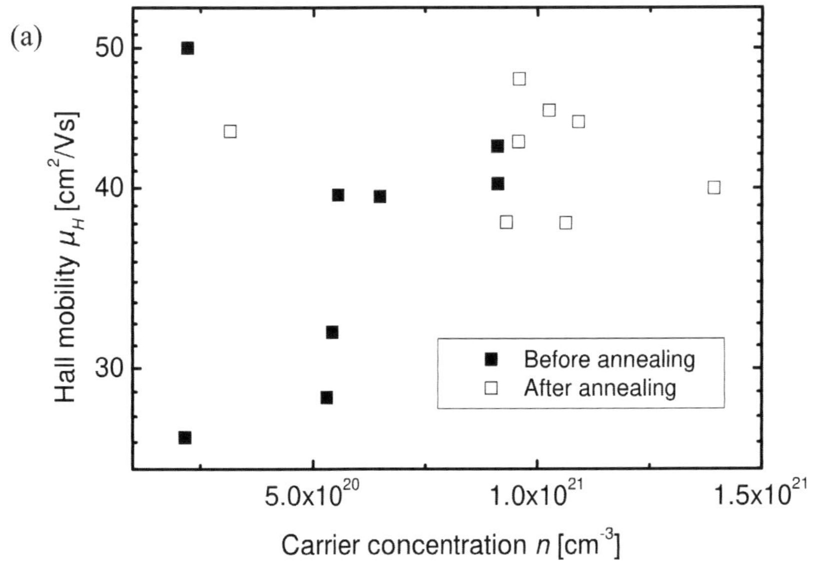

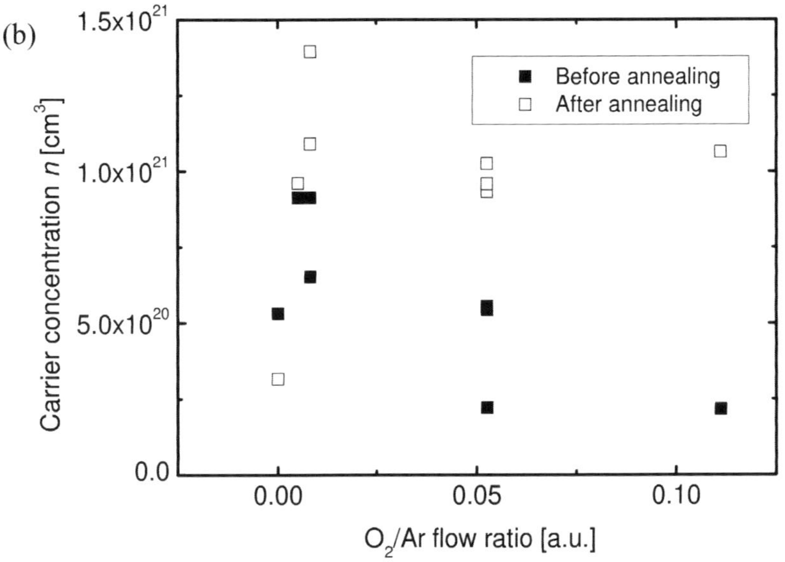

Figure 11. (a) Hall mobility in dependence to the carrier concentration for ITO thin films before and after annealing. (b): Carrier concentration as a function of flow ratio between oxygen and argon during deposition of ITO thin films.

Further we applied the same annealing procedure to ITO. Figure 11 (a) shows the mobility in dependence to the carrier concentration for ITO thin films before (full squares) and after (open squares) annealing, respectively. Similar to ZnO films we observed in most cases a strong decrease of resistivity upon annealing under the Si capping layer. However, ITO films show a quite different behavior upon annealing in terms of mobility and carrier density. As deposited ITO films reveal high mobilities between 25 cm^2/Vs and 50 cm^2/Vs and carrier concentrations in the range of 2 to 9 x 10^{20} cm^{-3}. Unlike the other ITO films, the film with the highest mobility shows degradation upon annealing down to 38 cm^2/Vs. For all other films an improvement in mobility is seen and all films exhibit mobilities between 38 cm^2/Vs and 48 cm^2/Vs after annealing. The highest increase was noticed for a sample with the initial value of 29 cm^2/Vs and 44 cm^2/Vs after annealing, respectively. The most remarkable effect, however, is the increase in carrier density, observed for most ITO films upon annealing. All films reveal carrier density close to or even above 10^{21} cm^{-3}.

Since the films were grown under different sputtering conditions and carrier density shows the most remarkable effect, Figure 11 (b) shows the dependence of the carrier concentration on the oxygen/argon flow ratio during deposition. It is seen that before annealing the carrier concentration generally exhibits a decrease as the oxygen flow increases, which is attributed to a reduced amount of ionized oxygen vacancies in the ITO films [20,21]. In general, almost all films, except the sample with no additional oxygen flow, show increase in carrier concentration upon annealing. Usually, this is explained by an out diffusion of oxygen atoms at high temperatures [20,21]. In our case, the films are protected by a silicon capping layer, so out-diffusion is limited. For ITO it is known, that annealing can cause crystallization of amorphous regions leading to a significant increase in carrier concentration [22]. The carrier density after annealing seems to be independent on the oxygen addition during sputtering, so activation of idle dopants at high temperatures is assumed to be the major effect to release free charge carriers to the lattice [23].

4. Conclusion

High temperature annealing was applied to several transparent oxides covered with an amorphous Si layer to protect them against degradation by environmental influences. The mobility of ZnO based films increased upon this annealing process and the highest mobility value of 72 cm^2/Vs was achieved for Al doped ZnO. The cap deposition and annealing process can easily be applied and hence could be utilized for large scale industrial applications. Ga as an alternative doping of ZnO did not change the annealing effect, i.e. an increase up to 59 cm^2/Vs was achieved. Further optimization is needed to reach similar values as for ZnO:Al. Different structural properties of ZnO films prepared by ion beam assisted deposition do not show clear tendency on mobility after annealing. This might be related to relaxation and healing of the defects during the annealing process. Incorporation of Mg into ZnO resulted in an increase of the band gap up to 3.7 eV of $Zn_{1-x}Mg_xO$:Ga films, but the electrical properties are deteriorated. Transmission measurements reveal increase in the transparency for all measured films, especially in the short wavelength and infrared range, except for sample $Zn_{0.97}Mg_{0.03}O$:Ga where slight deterioration after annealing was noticed. Increase in mobility and transmission makes these annealed films to suitable and competitive materials for front contact in thin film solar cells.

ITO films itself showed slight increase in mobility but the main effect was an increase in carrier concentration. Healing of intra-grain defects and defects at grain boundaries is assumed to be the major effect for ZnO leading to largely enhanced transport properties. ITO does not develop very high mobility upon annealing, but profits from an increase in carrier density. Therefore, in opposition to ZnO a capping layer is not necessarily needed for improvement in electrical properties of ITO films.

5. Acknowledgements

Thanks to our colleague Andreas Schmalen, Andreas Mück, Uwe Rau, Dominik Köhl from RWTH Aachen and Wilma Dewald from Fraunhofer IST, Braunschweig for technical support and fruitful discussions on NF_3 plasma etching, XRD and EPMA measurements, respectively. Ongoing financial support by the German Federal Environment Ministry (BMU, grants 0327693A and 0325299A) and the project "Grundlagen und Technologie für die Silizium-Dünnschichtphotovoltaik auf der Basis von Niedertemperaturprozessen" (grant 0327625) is gratefully acknowledged.

References

[1] Beyer W, Hüpkes J and Stiebig H, 2007 *Thin Solid Films* **516** 147
[2] Zhu H, Bunte E, Hüpkes J, Siekmann H and Huang S M, 2009 *Thin Solid Films* **517** 3161
[3] Kim J H, Jeon K A, Kim G H and Lee S Y, 2006 *Appl. Surf. Sci.* **252** 4834
[4] Minami T, 2005 *Semicond. Sci. Technol.* **20** 35
[5] Wei W, Jin C, Narayan J and Narayan R J, 2009 *Sol. Stat. Comm.* **149** 1670
[6] Maejima K, Shibata H, Tampo H, Matsubara K and Niki S, 2010 *Thin Solid Films*, **518** 2949
[7] Hüpkes J, Zhu H, Owen J I, Jost G, Bunte E, 2012 *Thin Solid. Films, **520*** 1913
[8] Berginski M, Hüpkes J, Schulte M, Schöpe G, Stiebig H and Rech B, 2007 *J. Appl. Phys.* **101** 074903
[9] Berginski M, Hüpkes J, Reetz W, Rech B and Wuttig M, 2008 *Thin Solid Films* **516** 5836
[10] Warzecha M, Hüpkes J, Doumit A, Köhl D and Owen J I, In *Proceedings of the 25th Eur. Photov.Sol. Energy Conf. and Exhib.*, Barcelona, Spain 2010, doi: 10.4229/25thEUPVSEC2010-3AV.1.16
[11] Köhl D, Luysberg M and Wuttig M, 2010 *J. Phys. D: Appl. Phys.* **43** 205301
[12] Ohtomo A, Kawasaki M, Koida T, Masubuchi K, Koinuma H, Sakurai T, Yoshida Y, Yasuda T and Segawa Y, 1998 *Appl. Phys. Lett.* **72** 2466
[13] Ruske F, Roczen M, Lee K, Wimmer M, Gall S, Hüpkes J, Hrunski D and Rech B, 2010 *J. Appl. Phys.* **107** *013708*
[14] Agashe C, Kluth O, Hüpkes J, Zastrow U, Rech B and Wuttig M, 2003 *J. Appl. Phys.* **95** 1911
[15] Van der Drift A, 1967 *Phillips Res. Rep.* **22** 267
[16] Wimmer M, Ruske F, Scherf S and Rech B, 2011 *Thin Solid Films*, doi: 10.1016/j.tsf.2011.04.102
[17] Seto J Y, 1975 *J. Appl. Phys.* **46** 5247
[18] Bachari E M, Baud G, Ben Amor S and Jacquet M, 1999 *Thin Solid Films* **348** 165
[19] Burstein E, 1954 *Phys. Rev.* **93**, 632
[20] Wen-Fa Wu and Bi-Shiou Chiou, 1996 *Sem. Sci. Technol.* **11** 196
[21] Hu Y, Diao X, Wang C, Hao W and Wang T, 2004 *Vacuum* **75** 183
[22] Morikawa H and Fujita M, 1999 *Thin Solid Films* **339** 309
[23] Yang C-H, Lee S-C, Chen S-C, Lin T-C, 2006 *Mat. Sci. Eng.* B **129** 154

Reduction of channel resistance in amorphous oxide thin-film transistors with buried layer

Eugene Chong[1, 2], Bosul Kim[2,3], and Sang Yeol Lee[4]

[1] Electronic Materials Research Center, Korea Institute of Science and Technology (KIST), Seoul, 136-791, Republic of Korea.

[2] Nanomaterials Science and Engineering, University of Science and Technology (UST), Daejeon, 305-333, Republic of Korea.

[3] Convergence Components & Materials Research Laboratory, Electronics and Telecommunications Research Institute (ETRI), Deajeon 305-700, Republic of Korea

[4] Department of Semiconductor Engineering, Cheongju University, Cheongju, Chungbuk, 360-764, Republic of Korea

E-mail: sylee@cju.ac.kr

Abstract. A silicon-indium-zinc-oxide (SIZO) thin film transistor (TFT) with low channel-resistance (R_{CH}) indium-zinc-oxide (In_2O_3:ZnO = 9:1) buried layer annealed at low temperature of 200°C exhibited high field-effect mobility (μ_{FE}) over 55.8 cm^2/V·s which is 5 times higher than that of the conventional TFTs due to small threshold voltage (V_{th}) change of 1.8 V under bias-temperature stress (BTS) condition for 420 minutes. The low-R_{CH} buried-layer allows more strong current-path formed in channel layer well within relatively high-R_{CH} channel-layer since it is less affected by the channel bulk and/or back interface trap with high carrier concentration.

1. Introduction

As a potential thin film transistor (TFT) alternative for current devices with low-cost, oxide thin film transistor with high-performance and low-temperature process (< 200°C) is gaining importance.[1-4] For next-generation displays, it requires high-speed driving circuits with TFTs to operate at low voltages and high driving currents as well as a small change of threshold voltage (V_{th}) during operation.[4-6] ZnO-based TFTs have been developed as the most possible candidates caused by their superb electrical properties especially In-Zn-O (IZO) system. However, it has been reported that there is a trade-off relationship between 1) process temperature and reliability and 2) mobility and stability of TFTs dependent on carrier suppressor (stabilizer) ratio, such as Ga, Zr, Hf, Mg and Si.[7-12] In other words, IZO TFTs without stabilizer/high temperature process up to 300 ± 50°C tend more easily to react with the oxygen/hydrogen molecule, showing unstable electrical properties at the back channel and/or interface between channel and gate insulator.[8-12] Therefore, it would be promising to seek channel candidates with better performance on plastic substrates. Also it is necessary to use suitable annealing process and reduce the channel resistance (R_{CH}) to obtain high-performance TFTs. Among many candidates of ZnO-based channel layer, Si-In-Zn-O is expected to be one of promising materials for low-temperature processing for flexible devices which requires specially low temperature process due to low melting point of flexible substrates.[3,7]

Published under licence by IOP Publishing Ltd

In this paper, we have focused on attaining a high performance SIZO-TFTs with low-temperature process below 200°C simultaneously by adopting the buried channel structure where a lower-resistance layer is embedded in the channel layer.[13,14] It is important to note that reducing the R_{CH} can increase the on-current (I_{ON}) of TFTs and field effect mobility (μ_{FE}).

2. Experimental Procedure

This paper has investigated the effect of low-resistance buried layer for 5SIZO TFTs with bottom-gate and top-electrode structure on the performance of buried oxide thin film transistor. Heavily doped p++-silicon wafers (0.01 Ω·cm), were used as gate electrode and substrates and SiO_2 as dielectric. To measure source-drain contact resistance (R_{SD}), transmission line method (TLM) patterns were fabricated on 5SIZO films and IZO-buried 5SIZO films using conventional etching and photolithographic process. The metal pad has dimensions of 300 μm × 100 μm and the pad-spacing/channel-length (l) of 50 to 250 μm. The TLM patterns had been used as TFT devices. The IZO-buried layer (IZO target; In_2O_3 : ZnO = 9:1 wt.%) and 5SIZO channel layer (5SIZO target; SiO_2 : In2O3 : ZnO = 5:71:24 wt.%) were deposited on SiO_2/p^{++}-Si substrates by rf-sputtering using rf-power of 45 W in Ar and Ar:O_2(97:3) ambient, respectively, at room temperature by fixing the total channel thickness at 45 nm. The base vacuum was lower than 1.0 × 10-6 Torr and working pressure is maintained 4.0 × 10^{-3} Torr while the active layer was deposited. The patterned channel layers are formed by conventional etching process using HCl (1 %). For all of devices, titanium (15 nm) and gold (60 nm) are deposited by e-beam and thermal evaporation methods, respectively, for contact electrodes. The electrodes are patterned by conventional photolithography and lift-off processes. Finally, the fabricated devices were annealed at 200°C in N_2 ambient for 1 h. The current-voltage (I-V) characteristics and electrical contact resistance were measured by semiconductor parameter analyzer (EL423, ELECS co.) at room temperature in a dark room.

3. Results and Discussion

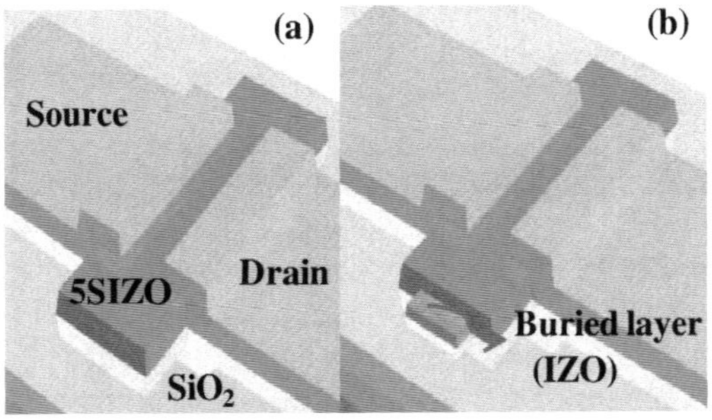

Figure 1. Schematic view of TFTs with (a) conventional single layer and (b) IZO-buried 5SIZO channel layer, respectively.

Figure 1(a) and (b) show the schematic diagram of conventional 5SIZO TFT and IZO-buried 5SIZO TFT structure, respectively. Because the electrical performance of the TFTs strongly relies on the R_{CH}, TFTs with different R_{CH} should exhibit different behaviour. Figure 2(a) and (b) show current- voltage (I-V) curves of conventional 5SIZO channel layer and 8 nm-thick IZO-buried 5SIZO channel layer measured at 5 V of pad-current by TLM as a function of l, respectively. It is clearly observed that the current level of suggested IZO-buried 5SIZO channel layer (Fig. 2(b)) is much higher than that of conventional 5SIZO channel layer (Fig. 2(a)), indicating that the IZO-buried 5SIZO channel layer for TFT device is expected to exhibit enhanced I_{ON} and μ_{FE} according to the equation [1],[15]

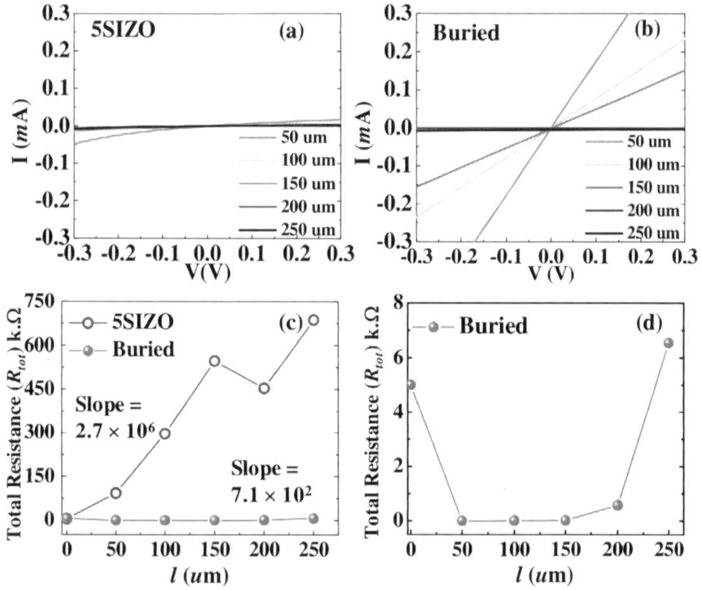

Figure 2. I-V curves of Ti/Au contact to (a) 5SIZO layer and (b) IZO-buried 5SIZO layer as a function of pad-spacing/channel length (l). (c) The $R_{tot} - l$ data for 5SIZO TFT and IZO-buried 5SIZO TFT annealed at 200°C. (d) Magnified $R_{tot} - l$ plot for IZO-buried 5SIZO TFT. The slope values of $R_{tot} - d$ for 5SIZO and IZO-buried TFTs are 2.7×10^6 and 7.1×10^2, respectively.

$$R_{CH} = l / [W\mu_{FE}C_i(V_{GS} - V_{th})] \quad [1]$$

where W is the channel width, l is the pad-spacing and channel-length, μ_{FE} is the field effect mobility, and C_i is the unit area capacitance of the dielectric. V_{GS} and V_{th} are the gate voltage and threshold voltage, respectively.

The μ_{FE} is calculated by linearly fitting the square root of I_{DS} vs. V_{GS} curve of the transistor operating in the linear region. The following is the expression for the operation of a field effect transistor in the linear region.[16]

$$\mu_{FE} = \frac{g_m l}{WC_i V_{DS}} \quad [2]$$

where W is the channel width, and g_m is the transconductance. Its subthreshold swing ($S.S$) was extracted from the subthreshold region of the linear portion of the log(I_{DS}) vs V_{GS} plot at I_{DS} of 0.1 V range between 10^{-9} and 10^{-10} A. Figure 2(c) shows the total resistance ($R_{tot} = I / V$) of the 5SIZO and IZO-buried TFTs as a function of l (TLM plots). The R_{tot} was determined from the slope of the linear region of the presaturated I_{DS} - V_{DS}, because, in the saturation region, the channel resistance becomes drain-voltage dependent.[17] The total resistance measured is a function of the contact resistances of the two contacts and the sheet resistance of the channel:[15,18]

$$R_{tot} = R_{CH} + 2R_C \quad [3]$$

where the R_{CH} is the channel resistance per unit l and the $l = 0$ intersection gives the contact resistance (R_C). The solid circles and open circles are for the TFTs with single 5SIZO and IZO-buried channel layers, respectively. From these plots, the R_{CH} of the TFTs with 5SIZO and IZO-buried channel layers

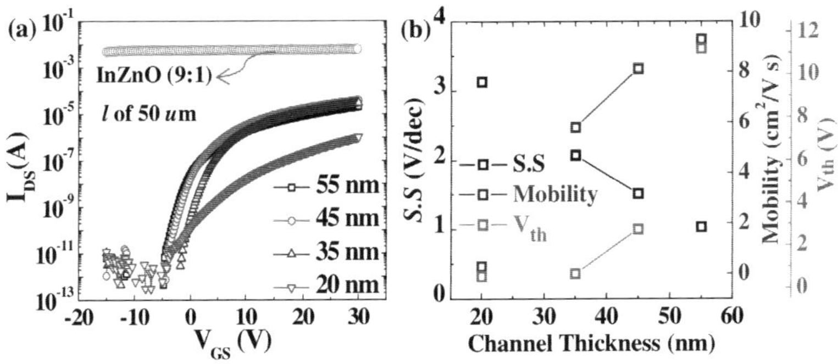

Figure 3. (a) Transfer curves of 5SIZO-TFTs for different channel thickness from 22 to 55 nm and IZO TFT and (b) extracted electrical properties from (a): subthreshold swing (*S.S*), field effect mobility (μ_{FE}) and threshold voltage (V_{th}) when *l* is 50 μm.

are derived to be about 1.35 × 10^8 and 3.5 × 10^4 Ω, respectively, whereas, the R_C between Ti/Au electrode and 5SIZO channel layer is estimated to be about 5.72 × 10^6 Ω from conventional 5SIZO TFTs. The increase of R_{CH} is likely due to the decrease of the absolute carrier number according to the number of initial carriers in the 5SIZO film that is smaller than the IZO-buried film.[19] Therefore, it is interesting to note that R_{CH} is significantly reduced by the order of 4 as we intentionally proposed due to low resistance of buried layer with the same R_C for both types of TFTs.
The comparison of the performance of 5SIZO TFTs with different channel thicknesses and IZO TFTs has been shown in Fig. 3 (a) and (b). As increasing the thickness of channel layer in 5SIZO TFTs, the I_{DS} level is increased to 3.9 × 10^5 A, μ_{FE} is increased to 9.27 $cm^2/V \cdot s$, and the V_{th} shifted in the negative direction to 0.55 V. Interestingly, the off-current of 5SIZO TFTs with channel thickness over 60 nm disappeared under similar I_{DS} level of ~ 10^5 A (data not shown here). Additionally, the employed IZO (In_2O_3:ZnO = 9:1 wt %) layer has high I_{ON} of ~ 10^{-2} A as measured by the semiconductor analyzer and the low surface resistance extracted from Fig. 3(a). This indicates that it is difficult to show TFT performance only with highly conductive IZO single channel layer.

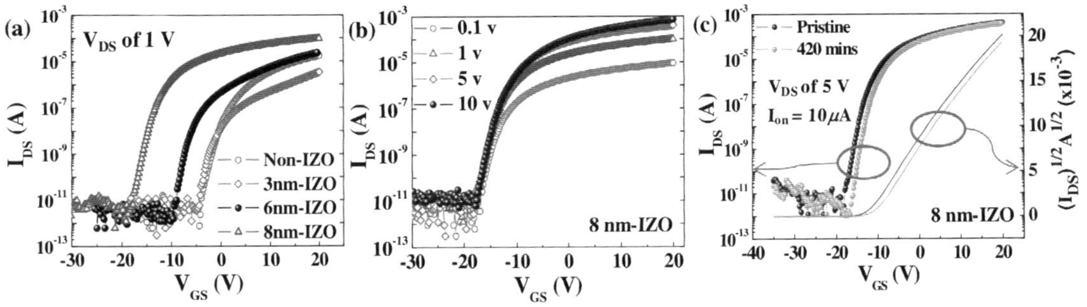

Figure 4. (a) The representative transfer curves of IZO-buried TFTs as a function of the thickness of buried-layer from 0 to 8 nm. (b) The transfer curve of optimized buried TFTs with 8 nm-IZO buried layer and (c) Transfer curves before and after BTS of 420 mins for buried TFT without passivation layer.

Figure 4 shows (a) representative transfer curves of IZO-buried TFTs as a function of the thickness of buried-layer from 0 to 8 nm. All devices had total 45 nm-thick channel layer. The TFTs with IZO-buried 5SIZO layer had high mobility and I_{ON} at V_{DS} of 1 V as shown in Fig. 4(a). We have systematically investigated the effect of the thickness of the buried-layer on the performance of TFTs since the conductance of buried-layer is much higher than that of channel layer. When the thickness of IZO-buried layer systematically increased from 0 to 8 nm (0, 3, 6 and 8 nm) in the 5SIZO TFTs, the

μ_{FE} increased from 8 to 55.8 $cm^2/V \cdot s$ (8, 11.6, 28.9 and 55.8 $cm^2/V \cdot s$) and V_{th} shifted from 0.67 to -11.9 V (0.67, -1.68, -4.93 and -11.9 V), respectively as a result of increasing low-resistance region and higher carrier concentration. As the low-resistance region is increased, the barrier between source and drain is lowered and it causes more injection of carriers. Figure 4(b) shows representative transfer curve of 8 nm thick IZO-buried 5SIZO TFTs at various V_{DS} values (0.1, 1, 5 and 10 V). These results clearly show I_{DS} saturation and increasing I_{ON} (from 9×10^{-6} to 7.2×10^{-4} A) indicating that the electron transportation in active channel layer is totally controlled by the gate and drain voltages. Also, it is very important to note, in terms of energy savings, that this IZO buried SIZO TFT is adjustable for low-power consumption device because this device can show excellent electrical properties even at low I_{DS} of 0.1 V.

Figure 4(c) shows the comparison of bias-temperature-stress (BTS) test result for the IZO-buried TFTs without any passivation treatment. Following the stress, the change of V_{th} was positive-shifted by about 1.8 V, which was measured at 60°C for 420 mins with on-current level of 10 μA at V_{DS} of 5 V. This (ΔV_{th} of 1.8 V) is comparable with ΔV_{th} of 2 V of conventional a-SIZO TFT with passivation in previous work[7], and it implies the channel layer protects buried-layer from the ambient effect.

4. Conclusion

We systematically demonstrate that the IZO-buried SIZO TFT shows excellent performances and bias-stability without any passivation treatment. This is achieved mainly because low-resistance buried-layer embedded between 5SIZO channel layer and gate insulator can clearly reduce the channel resistance. The R_{tot} for the buried-TFT is about 10^4 times lower than that for the conventional TFT at channel length 50 μm, mainly due to the higher carrier concentration of the IZO layer. Importantly, it has been observed that the μ_{FE} increased from 8 to 55.8 $cm^2/V \cdot s$ for the buried-TFTs having the same channel thickness of 45 nm. This is very important for the demands of future displays, such as 3-dimensinal (3D) and ultra-definition (UD) TV which requires high mobility more than 10 $cm^2/V \cdot s$. Therefore, the limitation of ZnO-based TFTs such as mobility and processing temperature can be resolved using a novel low-resistance buried-layer, especially low-temperature processed TFTs with SIZO channel layer.

Acknowledgments

This work was financially supported by a Korea Institute of Science and Technology (KIST) research program (Grant No. 2E22121).

References

[1] E. M. C. Fortunato, P. M. C. Barquinha, A. C. M. B. G. Pimentel, A. M. F. Goncalves, A. J. S. marques, L. M. N. Pereira and R. F. P. Martins 2005 *Adv. Mater.* **17**, 590.
[2] H. Frenzel, A. Lajn, H. v. Wenckstern, M. Lorenz, F. Schein, Z. Zhang, and M. Grundmann 2010 *Adv. Mater.* **20**, 5332.
[3] E. Chong, S. H. Kim, and S. Y. Lee 2010 *Appl. Phys. Lett*, **97**, 252112.
[4] J. Liu, D. B. Buchholz, R. P. H. Chang, A. Facchetti, and T. J. Marks 2010 *Adv. Mater.* **22**, 2333.
[5] L. Yuan, X. Zou, G. Fang, J. Wan, H. Zhou, and X. Zhao 2011 *IEEE Electron Devices Lett.* **32**, 42.
[6] T.-K. Kang, T.-C. Liao, C.-M. Lin, H.-W. Liu, and H.-C. Cheng 2011 *IEEE Electron Devices Lett.* **32**, 330.
[7] E. Chong, Y. S. Chun, and S. Y. Lee 2011 *Electrochem. Solid-State Lett.* **14**, H96.
[8] H. Yabuta, M. Sano, K. Abe, T. Aiba, T. Den, H. Kumomi, K. Nomura, T. Kamiya, and H. Hosono 2006 *Appl. Phys. Lett.* **89**, 112123.
[9] J.-S. Park, K.-S. Kim, Y.-G. Park, Y.-G. Mo, H.-D. Kim, and J.-K. Jeong 2009 *Adv. Mater.* (Weinheim, Ger.) **21**, 329.
[10] W.-H. Jeong, G.-H. Kim, H.-S. Shin, B.-D. Ahn, H.-J. Kim, M.-K. Ryu, K.-B. Park, J.-B. Seon, and S.-Y. Lee 2010 *Appl. Phys. Lett.* **96**, 093503.
[11] G.-H. Kim, W.-H. Jeong, B.-D. Ahn, H.-S. Shin, H.-J. Kim, H.-J. Kim, M.-K. Ryu, K.-B. Park, J.-B. Seon, and S.-Y. Lee 2010 *Appl. Phys. Lett.* **96**,163506.
[12] E. Chong, Y. S. Chun, S. H. Kim, and S. Y. Lee 2011 *Thin Solid Films.* **519**, 6881.
[13] E. Chong, Y. W. Jeon, Y. S. Chun, D. H. Kim, S. Y. Lee 2011 *Thin Solid Films.* **519**, 4347.
[14] F. Xue, H. Zhao, Y.-T. Chen, Y. Wang, F. Zhou, and J. C. Lee 2011 *Appl. Phys. Lett.* **98**, 082106.
[15] H. Kim, K.-K. Kim, S.-N. Lee, J.-H. Ryou, and R. D. Dupuis 2011 *Appl. Phys. Lett.* **98**, 112107.
[16] J. S. Park, J. K. Jeong, Y. G. Mo, H. D. Kim, and C. J. Kim 2008 *Appl. Phys. Lett.* **93**, 033513.
[17] M. Imakawa, K. Sawabe, Y. Yomogida, Y. Iwasa, and T. Takenobu 2011 *Appl. Phys. Lett.* **99**, 233301.
[18] S. Lee, H. Park, and D. C. Paine 2011 *J. Appl. Phys.* **109**, 063702.
[19] J. H. Noh, S. Y. Ryu, S. J. Jo, C. S. Kim, S.-W. Sohn, P. D. Rack, D.-J. Kim, and H. K. Baik 2010 *IEEE Electron Devices Lett.* **31**, 567.

IOP Conf. Series: Materials Science and Engineering **34** (2012) 012006 doi:10.1088/1757-899X/34/1/012006

Interface state density distribution in Au/n-ZnO nanorods Schottky diodes

S M Faraz[1, 2], M Willander[3] and Q Wahab[1, 2]
[1]Department of Electronic Engineering, NED University of Engineering and Technology, 75270, Karachi, Pakistan
[2]Departments of Physics, Chemistry and Biology, Linköpings University, SE-581 83, Linköping, Sweden
[3]Department of Science and Technology, Campus Norkoping, Linkoping University SE-581 83, Linköping, Sweden

Emai: sadia@ifm.liu.se

Abstract. Interface states density (N_{SS}) distribution is extracted in Au/ ZnO Schottky diodes. Nanorods of ZnO are grown on silver (Ag) using aqueous chemical growth (ACG) technique. Well aligned hexagonal–shaped vertical nanorods of a mean diameter of 300 - 450 nm and 1.3 - 1.9 μm high are revealed in SEM. Gold (Au) Schottky contacts of thickness 60 nm and 1.5mm diameter were evaporated. For electrical characterization of Schottky diodes current-voltage *(I-V)* and capacitance-Voltage *(C-V)* measurements are performed. The diodes exhibited a typical non-linear rectifying behavior with a barrier height of 0.62eV and ideality factor of 4.3. Possible reasons for low barrier height and high ideality factor have been addressed. Series resistance (R_S) has been calculated from forward I-V characteristics using Chueng's function. The density of interfacial states (N_{SS}) below the conduction band (E_C-E_{SS}) is extracted using *I-V* and *C-V* measured values. A decrease in interface states density (N_{SS}) is observed from 3.74 x 10^{11} – 7.98 x 10^{10} eV^{-1} cm^{-2} from 0.30eV - 0.61eV below the conduction band edge.

Keywords: Schottky contacts, Interface states, ZnO nanorods

1. Introduction

Zinc Oxide (ZnO) is a direct and wide band gap (3.37 eV) semiconductor with a large exciton binging energy (60 meV). It has attracted global interest in the research community due to its superior physical properties such as high electron saturation velocity, high breakdown electric field strength, high radiation tolerance and thermal conductivity. It exhibits near-ultraviolet (UV) emission which translates into bright light therefore it is famous for UV light emitters/detectors and high-power and high-temperature devices. [1, 2].Additionally it is piezo-electric, bio-safe and bio-compatible and has ferro-magnetic properties with potential for spintronics devices.

Schottky barrier diodes are the basis of a large number of compound semiconductor electronic devices such as field effect transistors, microwave diodes and photovoltaics.
High quality Schottky contacts are critical for realizing ZnO based devices [3, 4, 5]. Good quality metal–semiconductor interfaces, is always required, but is sensitive to the semiconductor surface, therefore understanding and realization of Schottky contacts is essential to continue the improvements in device performance.

Published under licence by IOP Publishing Ltd

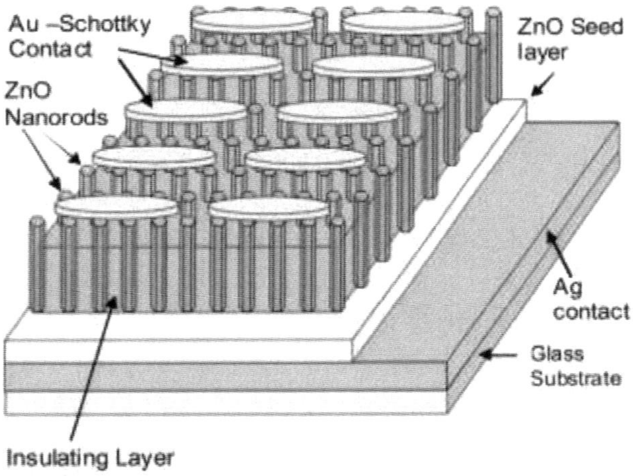

Figure 1 Schematic diagram of Au/ZnO nanorods Schottky diodes.

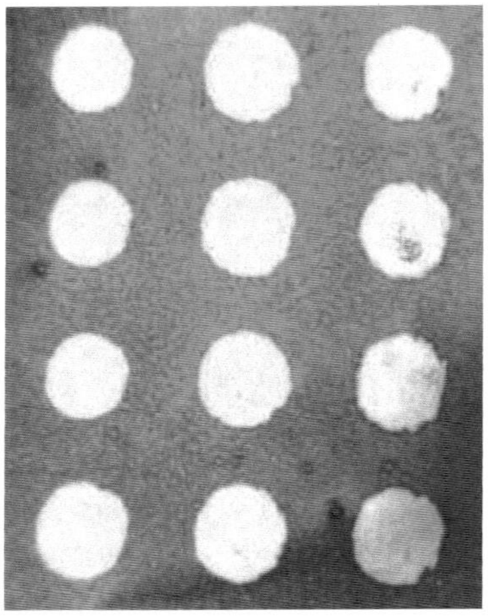

Figure 2 Au/ZnO Schottky diodes.

Interface states are a major factor which determines the Metal/ZnO contact characteristics. These states are generated due to surface termination, contaminations and metallization processes. Low temperature growth of ZnO nanorods by aquous chemical growth technique (ACG) may induce interface states to trap the carriers [6]. These states influence the charge transport mechanism and carrier recombination processes during the device operation and for optimal device performance the interface states should be below a certain level [7].

This paper deals with the study of the interface state distribution in Au/ZnO Schottky diodes. Schottky diodes are fabricated on ZnO nanorods and are characterized by current–voltage *(I–V)* and capacitance–voltage *(C–V)* measurement at room temperature. Diode parameters such as barrier height (φ_B), ideality factors (*n*) and series resistance (R_S) are extracted. The distribution of the interface state density (N_{SS}) within the band gap is investigated taking into account the voltage-dependent ideality factor *n* (V) and effective barrier heights (φ_e) obtained from *I–V* and *C–V* measured values. A correlation between high ideality factor with N_{SS} is also discussed

2. Experimental

ZnO nanorods have been grown on glass substrate by using aqueous chemical growth (ACG) method. Prior to the growth, the glass substrates were ultrasonically cleaned for 15 min with each acetone, methanol and de-ionized water. A layer of silver (Ag) of thickness 139nm has been evaporated on the cleaned glass substrate for ohmic contact [8]. In ACG growth technique an equimolar concentration of hexamethylenetetramine (HMT) ($C_6H_{12}N_4$) and zinc nitrate ($Zn(NO_3)_2 6H_2O$) 0.022–0.075 mM is used[9]. The substrate was placed in the solution and heated at a temperature of around 96 °C for about 4 hours. To improve the growth of nanorods, the substrate preparation has been done by the technique developed by Greene et al. [10]. Following this method a thin uniform seed layer has been spin coated on the substrate. Spin coating has been done three times to get uniform seed layer. Then the substrate was heated at 250 °C for 15 minutes. After the growth an insulating PMMA layer has been spin coated to fill the gaps between nano rods and to prevent the carrier cross talk amongst the nano-rods. It also isolates the electrical contacts on the nano-rods from reaching the substrate [11].

Oxygen plasma cleaning was performed to remove excess PMMA from the top of the nanorods. Gold (Au) circular Schottky contacts of 1.5mm diameter and 56nm thick were evaporated on to the top of the groups of nanorods in an ultra-high vacuum chamber through a shadow mask. The nanorods were characterized by using JEOLJSM-6301F scanning electron microscope (SEM). Current-voltage *(I-V)*

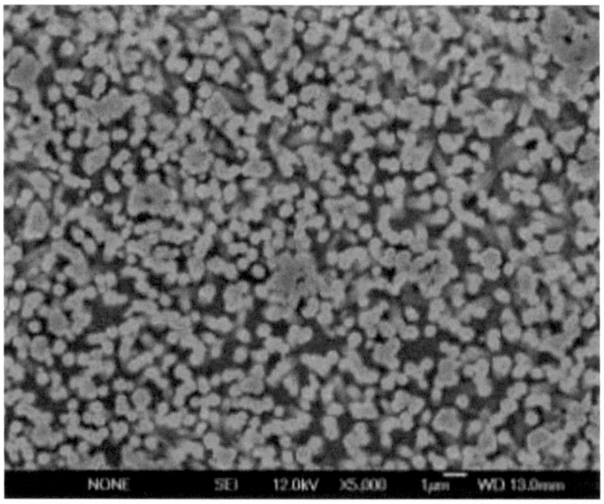

Figure 3 SEM image of ZnO nanorods.

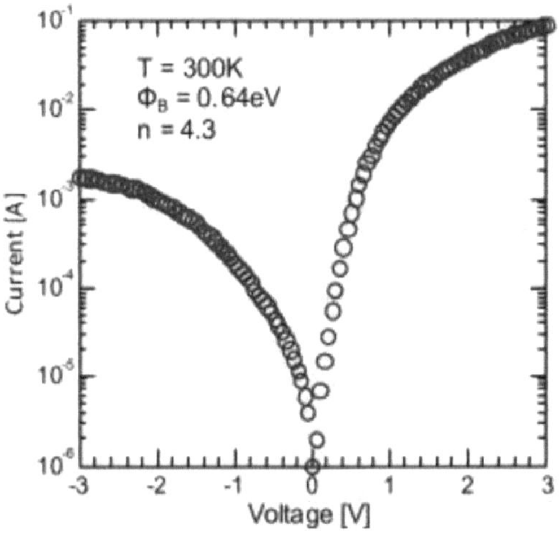

Figure 4 Current- votltage (I-*V)* characteristics of Au/ZnO Schottky

and capacitance-voltage *(C-V)* measurements have been performed using Keithley SCS-4200 by placing the sample on a probe-station. The Schematic diagram and actual image of Au/ZnO Schottky diode are shown in figure 1 & 2.

3. Results and Discussion

The size (diameter and height), distribution and morphology of the as grown nanorods have been studied by scanning electron microscopy (SEM) as shown in figure.3. Vertically well aligned hexagonal – shaped nanorods of around 1.3 – 1.9μm length and 300-450nm diameter have been revealed. The nanorods showed a tendency to grow perpendicular to the surface of substrate with a uniform distribution.

For electrical characterization current – voltage (*I-V*) and capacitance – voltage(*C-V*) measurements have been performed at room temperature. Measured forward and reverse *I-V* characteristics are shown in figure 4. Au/ZnO Schottky diodes showed a non – linear rectifying behavior with a barrier height of 0.62eV and ideality factor of 4.3 extracted using equation (1) & (2).

$$I_o = AA^*T^2 e^{-q\phi_B/kT} \tag{1}$$

$$I = I_o\left(e^{qV/nkT} - 1\right) \tag{2}$$

Here A^* is Richardson constant. For n-ZnO the theoretical value of Richardson constant used is 32 $A/cm^2\ K^2$ obtained from $A^* = 120 \times m_n^*/m_o$ $A/cm^2\ K^2$, where m_n^* is the effective mass of electrons which is $0.275m_o$ [12, 13]. A is the active area, which is estimated by taking into account the approximate nanorods density under the Schottky contact.

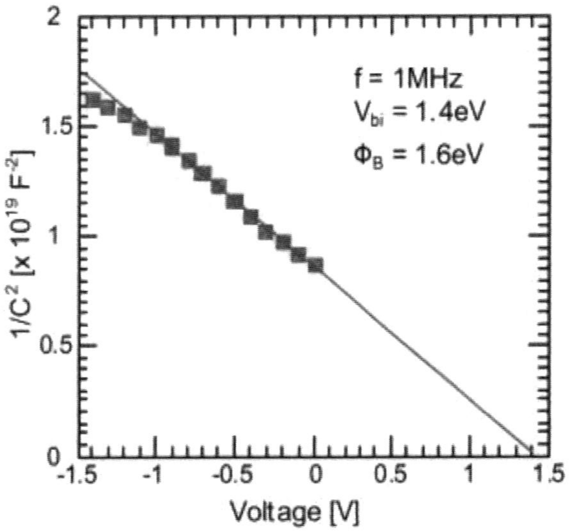

Figure 5 $1/C^2$ *vs* Reverse Voltage plot obtained at 1MHz

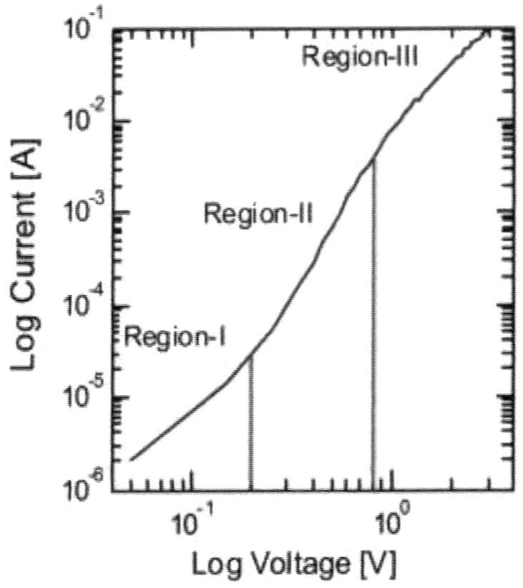

Figure 6 Forward bias logI - logV characteristics

The capacitance-voltage measurements have been performed up to 1 MHz and by plotting the inverse squared of the junction capacitance against the applied reverse voltage, the built-in potential (V_{bi}) and barrier height (φ_B) are obtained by a linear extrapolation of $1/C^2$ to the voltage axis (figure 5). The barrier height obtained from the *C–V* measurement is comparatively larger than the barrier heights obtained from the *I–V* measurements. It is due to the reason that sometimes the interface traps do not respond to applied ac signals and do not contribute to capacitance at higher frequencies [12].

High value of ideality factor obtained from I-V measurements, depict that the transport properties of diode cannot be well defined by thermionic emission only. To study the charge transport mechanism forward *logI - logV* characteristics have been studied. Room temperature *logI - logV* characteristics exhibited three distinct regions as shown in figure 6. At low voltages ($V < 0.2$) in region – I, an ohmic conduction is dominant and current varies almost linearly indicating that tunneling is the dominant charge transport mechanism. The second region shows exponential behavior of current ($0.2 < V < 0.8$) with ideality factors in the range of n = 2.3 - 3.7. It indicates that the transport mechanism is governed by recombination tunneling in this region. [14,15,16]. The current follows power law relation in region III which shows that space charge limiting current (SCLC) is the dominant charge transport mechanism. in region-III. The slope is measured to be 2.3 and the analysis of curve indicates that SCLC is might be controlled by the trap levels at different energies within the band gap of ZnO. The SCLC and the other regions observed in the present study are also reported by P.Klasson *et al* [15] and G.Amin *et al* [17] for Schottky contacts on ZnO nanorods.

For Schottky diodes the non-idealities are often observed due to various mechanisms such as tunneling, recombination current, interface states between the metal and the semiconductor and rough semiconductor surface [18]. At high forward bias the nonlinearities are also attributed to edge effects and series resistance (R_S). The series resistance (R_S) has been extracted using the method proposed by Chueng & Chueng [19] from the following equation,

$$\frac{dV}{d(\ln I)} = \frac{nkT}{q} + IR_S \tag{3}$$

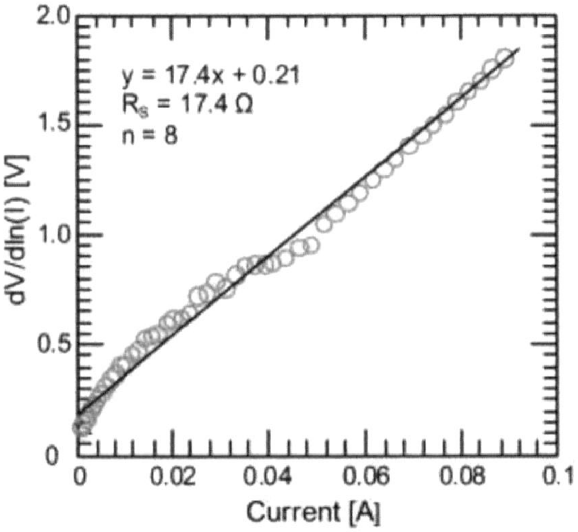

Figure 7 Plot of dV/d(lnI) vs. I

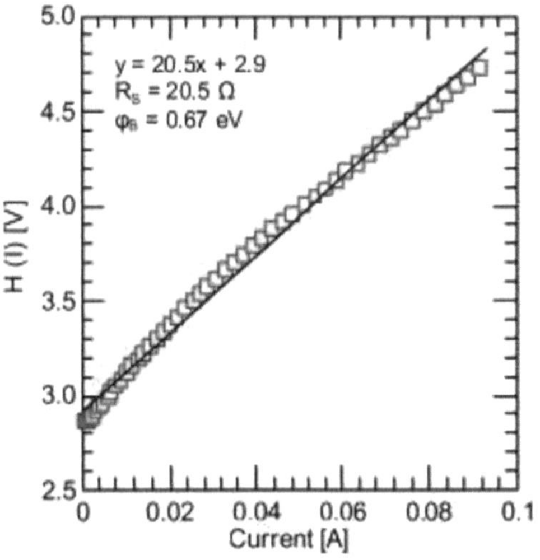

Figure 8 Plot of H(I) vs. I

The plot of *dV/d(lnI)* versus current is a straight line as shown in figure 7. The slope of the line and y-intercept give the values of the R_S and ideality factor (n). Series resistance and ideality factor are found to be 17.4 Ω and 8 respectively. Series resistance and barrier height are also calculated using the function H (I), given by equation (4) and (5).

$$H(I) = V - \left(\frac{nkT}{q}\right)\ln\left(\frac{I}{AA^{*}T^{2}}\right) \tag{4}$$

$$H(I) = n\phi_B + IR_S \tag{5}$$

From the plot of H(I) versus current, the slope and y-intercept of the straight line give Rs of 20.5 Ω and barrier height of 0.67eV (figure 8). Series resistances obtained from both plots are in close agreement with each other. The difference in the ideality factors (n) obtained from dV/dln(I) vs I plot and logI-V plot can be attributed to the series resistance, interface states and voltage drop across interfacial layer between Au and ZnO.

In fact interface states are always introduced due to oxidation of semiconductor surfaces from extended air exposure, surface dipoles, incomplete covalent bonds, hydrogen, chemical reactions during fabrication processes and sharp discontinuity between semiconductor crystal and metal [20, 21] Such defects produce a large density of interface states which are continuously distributed in the energy within the forbidden gap and cause leakage currents to flow. Taking into account the series resistance and voltage dependent ideality factor *n(V)* the density of interface states (N_{SS}) is extracted using I-V and C-V measured values using equation (6) [22].

$$N_{SS} = \frac{1}{q}\left[\frac{\varepsilon_i}{\delta}\{n(V) - 1\} - \frac{\varepsilon_S}{W_D}\right] \tag{6}$$

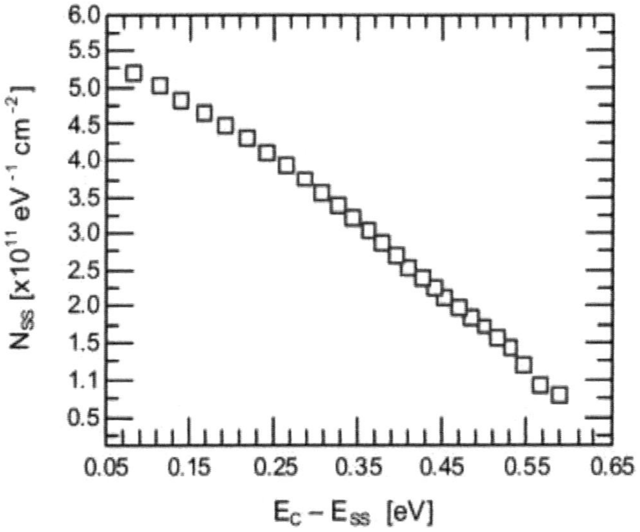

Figure9 Distribution of interface states (Nss) for Au/ZnO nanorods Schottky diode for ϕ_B= 0.62eV.

Here W_D is the width of the space charge region extracted from experimental *C-V* data [23], ε_I and δ are permittivity and thickness of the interfacial layer. The capacitance of interfacial layer is calculated from C-V measurements at 1MHz using equation (7).

$$C_i = C_{ma}\left[1+\left(\frac{G_{ma}}{\omega C_{ma}}\right)^2\right] \tag{7}$$

$$\frac{C_i}{A} = \frac{\varepsilon_i \varepsilon_o}{\delta} \tag{8}$$

Voltage dependent ideality factor n (V) has been extracted from equation (8).

$$n(V) = \frac{qV}{kT \ln\left(\frac{I}{I_o}\right)} \tag{9}$$

Bias dependant effective barrier height (φ_e) and the energy of interface states with respect to the conduction band edge (E_C-E_{SS}) are obtained by using equation (10) and (11) for the barrier height of 0.62 eV obtained from I-V measurements.

$$\phi_e = \varphi_{bo} + \left(1 - \frac{1}{n(V)}\right)(V - IR_S) \tag{10}$$

$$E_c - E_{SS} = q(\phi_e - V) \tag{11}$$

A decrease in interface states density (N_{SS}) is observed from 5.20 x 10^{11} – 7.98 x 10^{10} eV^{-1} cm^{-2} with an increase in Ec-Ess from 0.08eV - 0.59eV below the conduction band edge as shown in figure 9.

These surface states are formed in ZnO nanorods because of their large surface area and small diameter [24]. When electrons are excited to conduction band they are trapped by these states and hence the electrical behavior of diode is affected. The effects of interface states are more important for compound semiconductor nano structure as their surfaces have more defects [25].

4. Conclusion:

Au/ZnO Schottky diodes have been fabricated and characterized. The density of interface states N_{SS} have been extracted from I-V and C-V measured values. A decrease in the interface states density have been observed from 5.20 x 10^{11} – 7.98 x 10^{10} eV^{-1} cm^{-2} from 0.08eV - 0.59eV below the conduction band edge. The results yield that the interface states play a very important role in the current flow mechanism in electronic devices and these states must be kept as low as possible in order to reduce the surface recombination and tunneling.

References:

[1] Look D C 2001 *Mater. Sci. Eng: B* 80 383
[2] Wenckstern H V, Brandt M, Schmidt H, Biehne G, Pickenhain R, Hochmuth H, Lorenz M, and Grundmann M 2007 *Appl. Phys. A*, **88**, 135
[3] Mosbacker H L, Strzhemechny Y M, White B D, Smith P E, Look D C, Reynolds D C, Litton C W, and Brillson L J 2005 *Appl. Phy. Lett.* **87**,012102
[4] Pearton S J, Norton D P, Ip K, Heo Y W and Steiner T 2005 *Prog. Mater. Sci.* **50** 293
[5] Sarpatwari K, Awadelkarim O O, Allen M w, Durbin S M and Mohney S E 2009 *Appl. Phys. Lett.* **94** 242110
[6] Liou S C, Hsiao C S and Chen S Y 2004 *J. Crys. Growth* **274** 438
[7] Brillson L J, Mosbacker H L, Hetzer M J, Strzhemechhy Y, Jessen G H, Look D C, Cantwell G, Zhang J and Song J J, 2007 *Appl. Phys. Lett.***90** 102116
[8] Chaabouni F, Abaab M and Rezig B 2006 *Superlattices and microstructures* **39** 171
[9] Vayssieres L, Keis K, Lindquist S E and Hagfeldt A 2001 *J. Phys. Chem. B* **105** 3350
[10] Greene L E, Law M, Tan D H, Montano M, Goldberger J, Somorjai G, Yang P 2005 *Nano Lett.* **05** 1231
[11] Alvi N H, Willander M and Nur O 2010 *Superlattices and Microstructures* **47** 754
[12] Schroder D K 2006 Semiconductor material and device characterization 3rd ed. Wiley, New York
[13] Pra M, Csaba G, Erlen C and Lugli P 2008 *J. Comput. Electron.* **07** 146
[14] Reddy N K, Ahsanulhaq Q, Kim J H and Hahn Y B 2008 *Appl. Phys. Lett.* **92** 3127
[15] Klasson P, Nur O and Willander M 2008 *Nanotechnology* **19** 475202
[16] Kim D C, Han W S, Cho H K, Kong B H and Kim H S 2007 *Appl. Phys. Lett.* **91** 1901
[17] Amin G, Hussain I, Zaman S, Bano N, Nur O and Willander M 2010 *Phys. Status Solidi A* **207** 748
[18] Faraz S M, Ashraf H, Arshad M I, Hageman P R, Asghar M and Wahab Q 2010 *Semicond. Sci. Technol.* **25** 095008
[19] Chueng S. K and Chueng N. W. 1986 *Appl. Phys. Lett.* **49** 85
[20] Streetman B G and Banerjee S K 2006 *Solid State Electronic Devices* 6th edn (Englewood Cliffs, NJ: Prentice-Hall)
[21] Brillson L J 1983 *J. Phys. Chem. Solids* **44** 703
[22] Card H C and Rhoderick E H 1971 *J. Phys. D: Appl. Phys.* **04** 1589
[23] Sze S M 2006 *Physics of Semiconductor Devices* 3rd edn (New York: Wiley)
[24] Liu X, Wu X, Cao H and Chang R P H 2004 *J. Appl. Phys.* **95** 3141
[25] Sochackia M, Szmidta J,Bakowskib M, Werbowya A 2002 *Diamond and Related Materials* **11** 1263–1267

Al and Si doping of sputtered ZnO thin films

R. Schifano*, M. Schofield*, L. Vines*, S. Diplas, E. V. Monakhov*, and B. G. Svensson***

*) Department of Physics /Center for Material Science and Nanotechnology, University of Oslo, P.O. Box 1048 Blindern, N-0316 Oslo, Norway **) SINTEF Materials and Chemistry, P.O. Box 124 Blindern, Forskingsveien 1, N-0314 Oslo, Norway

E-mail: ramon.schifano@smn.uio.no

Abstract. In this work the effect of Al and Si content on the electrical, optical and structural characteristics of Radio Frequency (RF) sputtered ZnO thin films ($\sim 150\ nm$ thick) have been investigated. ZnO:Al and ZnO:Si films were deposited on glass through ZnO and Al targets and ZnO and Si targets co-sputtering, respectively. The Al and Si content have been varied between $\sim 10^{18}\ cm^{-3}$ and $\sim 4 \times 10^{21}\ cm^{-3}$, as determined by secondary ion mass spectrometry ($SIMS$) and X-ray photoelectron spectroscopy (XPS) measurements. The X-ray analysis, in the $\Theta - 2\Theta$ configuration, revealed a surfactant action of both Si and Al in the films, i.e., a reduction of the absolute stress $\perp$ to the c axis from $\sim 0.6 - 0.8\ GPa$ to $\sim 0.4\ GPa$ and $\sim 0.2\ GPa$ for Al and Si doping, respectively. In addition, a corresponding decrease of the root mean square (RMS) of the surface roughness with higher Al and Si content has been observed by Atomic Force Micropscope (AFM) measurements. Finally, the electrical characterization of the ZnO films, performed by room temperature Hall measurements, suggested that the best structural conditions in terms of grain size and absolute stress correspond, as expected, to the minimum resistivity.

1. Introduction

Recently, ZnO has been attracting increasing interest due to its unique combination of properties that makes it a promising material for the realization, among others, of solid state ultraviolet (UV) emitters, transparent and high power electronic devices [1, 2, 3]. In addition, the abundance of Zn makes ZnO one of the most promising candidates among the Transparent Conductive Oxides (TCOs) for large scale application like windows films for solar cell and displays. However, despite the wide research on this material and that most of the $CuIn_{1-x}Ga_xSe_2$ (CIS and CIGS) based photovoltaic modules already are based on ZnO as a conducting front contact the role of commonly used ZnO dopants, particularly in extremely high concentrations suitable for TCOs applications, is not fully clarified. As an example, the formation of dopant oxides and creation of intrinsic defects actually providing the free carriers has been proposed, but it is still under debate [4]. In addition, most of the low resistivity/high uniformity films deposited by magnetron sputtering have been accomplished by placing the substrates perpendicular to the targets and/or by performing the film depositon under a static magnetic field. However, the physics and technology behind these deposition conditions have not been fully resolved [5, 6, 7]. Hence a deeper understanding of the doping mechanisms and the sputtering dynamics is likely to enable preparation of ZnO based films with resistivities below the present limit of $\sim 2 \times 10^{-4}\ \Omega \cdot cm$ [7].

Published under licence by IOP Publishing Ltd

In contrast to Al-doped ZnO films, no detailed study of the structural and morphological characteristics of Si-doped ZnO films has been reported previously in the literature [8, 9, 10]. The purpose of this work is to compare the effects of Al and Si doping on the electrical and structural properties of ZnO films prepared by RF sputtering in an ordinary target versus substrate face to face configuration.

2. Experimetal details

The sputtering procedure and conditions used were based on work performed in order to optimize the crystal structure of not intentionally doped ZnO [11] and is summarized hereafter. All ZnO films studied in the present work were sputtered on ordinary glass. The glass substrates were cleaned in aceton, ethanol and deionized water for 5 min. sequentially, before being loaded into a Semicore balanced magnetron sputtering deposition chamber that accomodates two 3 inches water cooled RF powered cathodes and one 3 inches water cooled Direct Current (DC) powered cathode. A 99.99% pure ZnO ceramic target, a Si wafer and an 99.999% Al target were mounted in the two RF and the DC cathodes, respectively. Prior to deposition, the sample chamber was evacuated to a base pressure lower than $2 \cdot 10^{-6}$ $Torr$ after reaching the desired deposition substrate temperature and 99.999% pure Ar was introduced with a pressure of 16.5 ± 0.2 $mTorr$. The Si and Al doping was realized by co-sputtering and varying either the Al or Si cathode power in the $6-9$ W and $1-20$ W range, respectively, while keeping the ZnO cathode power between $48-50$ W. A constant substrate temperature of 400 ^oC and 450 ^oC for the Al and Si doped samples, as well as a rotation of the sample holder of 12 rpm, was maintained during the depostions.
The Al and Si impurity content have been monitored by $SIMS$ and XPS. The $SIMS$ characterization was performed using a Cameca IMS7f microanalyzer to extract the concentration versus depth profiles. Crater depths were measured with a Dektak 8 stylus profilometer, and the erosion rate was assumed to be constant when converting sputtering time to sample depth. Concentration calibration of Al and Si were performed using implanted reference samples with a Al and Si peak concentration of $\sim 6 \times 10^{18}$ cm^{-3}. The XPS measurements were performed using monochromatic Al Kα radiation ($h\nu = 1486.6$ eV) on a Kratos Axis Ultra^{DLD} instrument at a 0° angle of emission ($a.o.e.$). The comparison between the integrated intensity of the Al and Si $2p$ lines and the sum of the O $1s$, Zn $2p$ and the C based surface contaminants lines, was used to determine the Al and Si at. %, respectively.
Structural properties such as grain orientation, grain size and residual uniform stress were studied using a Siemens D5000 X-ray diffractometer (XRD) with locked-coupled $\theta - 2\theta$ scans in the $20^o - 80^o$ range. The root mean square roughness (RMS) was determined by Atomic Force Microscopy (AFM) measurements using a Veeco D3100 microscope running in tapping mode; five 1×1 μm images were taken and analyzed for each sample.
The resistivity was determined at room temperature with a LakeShore 7704A Hall system using a van der Pauw geometry. Finally, the thicknesses of the films were measured using a Rudolph Research Auto-EL ellipsometer or a Dektak 8 stylus profilometer after chemically etching steps.

3. Results analysis and discussion

3.1. Film structure, surface morphology and resisitvity

Figs.1 and 2 show the Al and Si content of the sputtered films versus the ratio of the power applied to the Al, Si and ZnO targets, respectively, as determined by $SIMS$ and XPS measurements. The samples with ~ 1 at. %, representing the upper and lower limit for the $SIMS$ and XPS measurements, respectively, have been analysed with both techniques to verify the validity of the concentration calibration used over the whole range of applied powers. Except

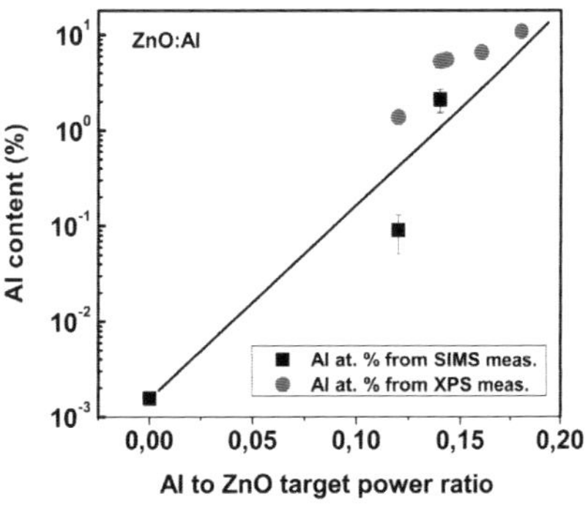

Figure 1. Al atomic percentage content of the sputtered films as determined by *SIMS* and *XPS* measurements versus the ratio of the power applied to the Al and ZnO target. The solid line is included as a guide to the eye and a linear dependence may not necessarily be valid.

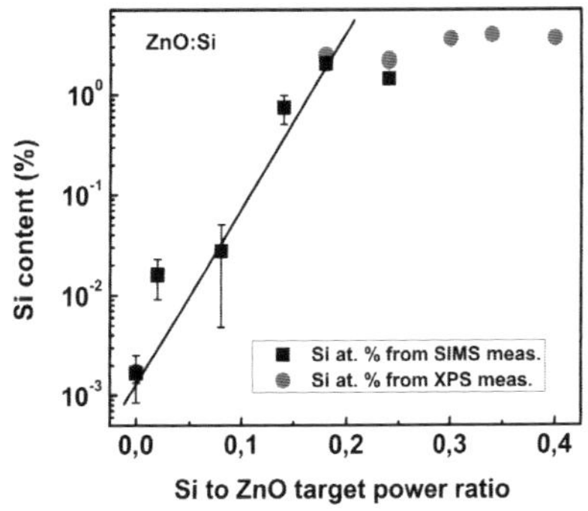

Figure 2. Si atomic percentage content of the sputtered films as determined by *SIMS* and *XPS* measurements versus the ratio of the power applied to the Si and ZnO target. The solid line indicates the regime where a linear dependence between the Si content and the Si to ZnO applied power ratio is observed, while above 9 W (0.18 in Si to ZnO power ratio) the Si content saturates at $\sim 3-4$ at. %.

for the ZnO:Al films deposited with a Al/ZnO power ratio of 0.12, the agreement between *SIMS* and *XPS* is reasonably good (within a factor of two) and in the following average values have been used. The reason for the large deviation at the power ratio of 0.12 is not fully clear, but it can be noted that the amount of incorporated Al is close to (or even below according to the *SIMS* data) the detection limit of *XPS*, implying a rather high uncertainty. Further, the area probed by *XPS* is $\sim$ 100 times larger than that by *SIMS*, but a systematic *SIMS*-mapping of the Al content over the surface did not reveal any non-uniformity. Hence, precipitation and formation of Al clusters can be ruled out as a cause for discrepancy at the power ratio of 0.12. Moreover, the solid line in Fig.1 is included as a guide to the eye only; the amount of data is not sufficient to claim a linear dependence and regions with different slopes may indeed exist. For the ZnO:Si samples, the *XPS* data reveal a saturation in the amount of incorporated Si atoms for power ratios above 0.18 (Si cathode powers $> 9\ W$). The probe depth of *XPS* at the conditions used (0° *a.o.e.*) is only $\sim 30\ \AA$ [12], thus mainly Si atoms incorporated as dopants will be monitored unless Si precipitates and/or nanocrystal are formed in the immediate vicinity of the surface. Hence, this possibly indicates that the maximum solid solubility of Si in ZnO is $\sim$ 2 at., in agreement with that Si nanocrystals embedded in a ZnO matrix form already for a Si at. % equal to $\sim$ 0.5 [10]. Further work is being pursued to confirm this scenario.

The analysis of crystal structure of the ZnO:Al and ZnO:Si films by *XRD* $\theta - 2\theta$ scans in the $20^o - 80^o$ range revealed only the presence of the (002) peak and its replica (004) (not shown), confirming the expected [13] preferred self-texturization of the films along the $< 0001 >$ direction and the absence of secondary phases like Zn, Al, Si and Al_2O_3 in the crystalline regions of the films.

However, the (002) peak angular position shifted and the *FWHM* varied with the Al and Si content indicating an influence on the mechanical stress and the grain size of the films, Fig.3

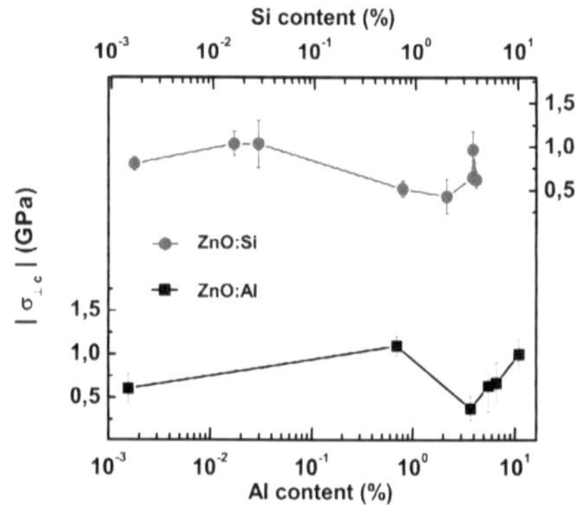

Figure 3. The total absolute in plane stress $|\sigma_{\perp c}|$ versus the Si and Al content.

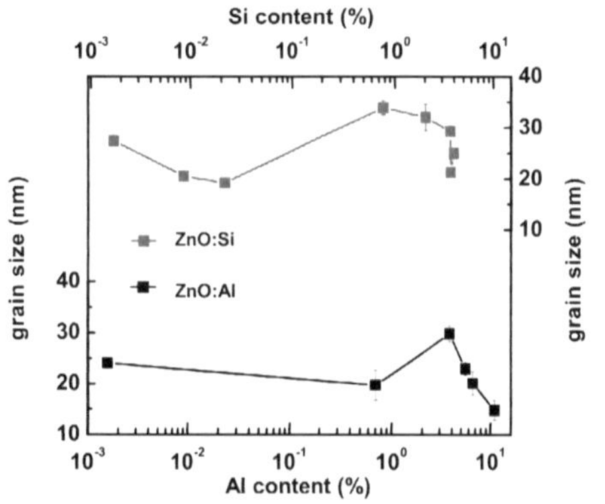

Figure 4. Grain size, $< L >$, versus the Si and Al content.

and Fig.4. Assuming the stress being uniformly distributed throughout the films, the absolute value of total component perpendicular to the growth direction $|\sigma_{\perp c}|$, i.e. the sum of the σ_{xx} and σ_{yy} stress components along the in plane x and y axis, has been evaluated according to the formula valid for hexagonal lattices [14]:

$$|\sigma_{\perp c}| = |\sigma_{xx} + \sigma_{yy}| = \left| \frac{2 \cdot C_{13}^2 - C_{33} \cdot (C_{11} + C_{12})}{2 \cdot C_{13}} \cdot \frac{c_{film} - c_{bulk}}{c_{bulk}} \right| \tag{1}$$

where c_{film} is the film and c_{bulk} the fully relaxed lattice constant in the c direction, respectively, the latter taken equal to ($d = 5.206 \pm 0.002$) Å [2]. For the elastic constants the nominal data for crystalline ZnO, suggested in Ref. [13], have been used: $C_{11} = 208.8\ GPa$, $C_{12} = 119.7\ GPa$, $C_{13} = 104.2\ GPa$ and $C_{33} = 213.8\ GPa$.

The XRD analysis revealed, as shown in Fig.3, a reduction of the absolute stress $\perp$ to the c axis from $\sim 0.6 - 0.8\ GPa$ for the not intentionally doped ZnO films to $\sim 0.4\ GPa$ and $\sim 0.2\ GPa$ when the Al and Si doping is equal to $\sim 2 - 4$ at. %. Furthermore, Fig.3 clearly indicates an increase in the amount of stress when the Al and Si content is either increased or decreased from these optimal values.

The grain size along the growth direction, $< L >$, is shown in Fig.4 and has been determined, assuming a Gaussian shape for the prominent (002) peak, according to the Scherrer formula [15]:

$$< L >= \lambda_{Cu_{K\alpha 1}} \cdot \frac{0.9394}{(\beta \cdot cos(2\Theta_{max}/2))} \tag{2}$$

with

$$\beta = \sqrt{\beta_{meas}^2 - \beta_{inst}^2} \tag{3}$$

where $2\Theta_{max}$ is the angle on the 2Θ scale where the (002) peak is observed, β_{meas} and β_{inst} are the $FWHM$ of the measured ZnO (002) and the Si (111) peaks, respectively, the latter taken as the Gaussian instrumental contribution to the overall $FWHM$. Note, that the $FWHM$ is also influenced by the shape distributions and any distortion of the crystal structure like microstrain, stacking faults, dislocations and twin planes. Such effects are not investigated in the present study and are, therefore, not be taken into account.

As shown in Fig.4, neglegting the above mentioned effects, it is found that also $< L >$ is affected by the presence of Al and Si with an increase in the grain size from $\sim 25\ nm$ for the not intentionally doped ZnO films to $\sim 30\ nm$ when the Al and Si doping is equal to ~ 4 at.

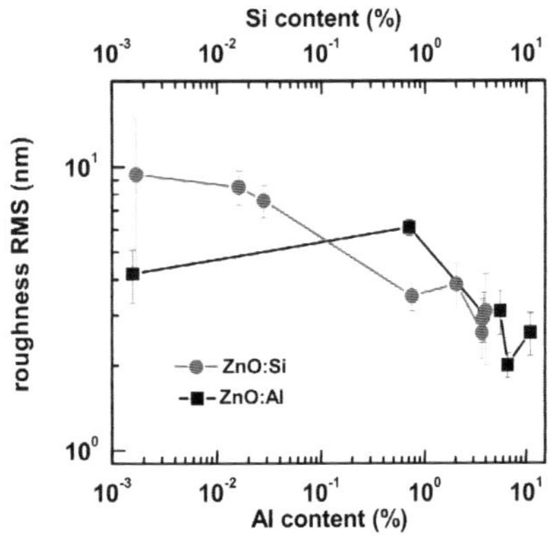

Figure 5. *RMS* of surface roughness versus the Al and Si content.

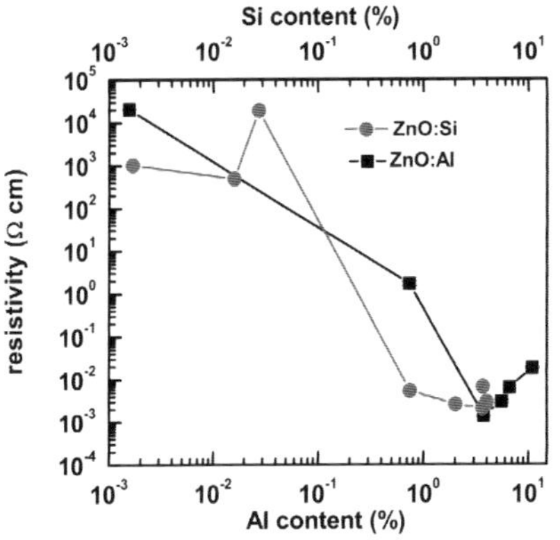

Figure 6. Electrical resistivity versus the Al and Si content.

% and ~ 0.8 at. %, respectively; that represents again roughly the optimal values with a trend versus lower $< L >$ when the Al and Si content is varied.

The dependence of the *RMS* surface roughness on the Al and Si content, as determined by the *AFM* analysis, is shown in Fig.5. The *RMS* values are found to depend strongly on the portion of the sample surface investigated, as represented by the large error bars. Similar results, have been reported by Minami et al. [16] for ZnO:Al films and they found a correlation with erosion patterns (preferential sputtering) of the ZnO target; it was speculated that high energy O atoms impinge on the growing films and cause local damage of the crystal structure. Indeed, elevated substrate temperature ($\gtrsim$ 350 ^oC) reduces the structural inhomogeneitis (and the scatter in the electrical resisitivity) [17], but our data indicate that temperatures of 400 - 450 ^oC are not sufficent to eliminate the effect on the *RMS* surface roughness fully. However, despite the large error bars, a clear trend with decreasing *RMS* vs. higher Al and Si content exists.

These results point to the role of the dopants Al and Si as a mineralizer or surfactant, that significantly improves the texture of the ZnO layers and while this has been already reported in the case of Al [4, 13, 18, 19] it is shown here that Si doping has a similar effect in promoting the wetting of the substrate surface and improving the overall crystal quality of the films.

The electrical resistivity as a function of the Al and Si content is shown in Fig.6. For both Al and Si, a decrease by up to 7 orders of magnitude occurs from $\sim 1 - 10\ k\Omega \cdot cm$ down to $(1.41 \pm 0.05) \times 10^{-3}\ \Omega \cdot cm$ and $(2.1 \pm 0.1) \times 10^{-3}\ \Omega \cdot cm$ respectively, with the minimum occurring for a Al and Si at. % equal to (4 ± 2) and (3.6 ± 0.4). The optimum electrical characteristics occur at about the Al and Si at. % corresponding to the largest grain size and most relaxed films and hence, the electrical data confirm the findings of the film structural analysis. Our minimum resistivity values are about a factor two higher than the best ones reported in the literature ($\sim 5 \times 10^{-4}\ \Omega \cdot cm$) for ZnO:Al and ZnO:Si deposited on corning glass [8, 20].

4. Conclusions

In the present work it has been shown that Si acts similarly to Al on the electrical and structural properties of ZnO when embedded in the ZnO matrix. While the donor effect of Si in ZnO has already been reported its beneficial effect on the crystal quality, that is its surfactant action in ZnO, has not been sufficently underlined so far. Similarly to Al, the optimal Si content for achieving the best electrical and structural properties for TCO applications is found to occur within the $\sim 1 - 2$ at. % interval. Furthermore, it has been found that the optimal Si at. %

content is close to the solid solubility in polycrystalline ZnO (equal to $\sim 2-4$ at. %). However, additional work is needed to unambiguously confirm this relation.

Acknowledgments

This work has been performed within "The Norwegian Research Centre for Solar Cell Technology"project number 193829, a Centre for Enviroment-friendly Energy Research co-sponsored by the Norwegian Research Council and research and industry partners in Norway and the Frienergi program.

References to printed journal articles

[1] Jagadish C and Pearton S J 2006 *Zinc Oxide Bulk, Thin films and Nanostructures* (Elsevier, Oxford)
[2] Özgür, Alivov Y I, Liu C, Teke A, Reshchikov M A, Doğan S, Avrutin V, Cho S J and Morkoç H 2005 *J. Appl. Phys.* **98** 041301
[3] Pearton S, Norton D, Ip K, Heo Y and Steiner T 2005 *Prog. Mater. Sci.* **50** 293
[4] Ellmer K 2000 *J. Phys. D: Appl. Phys.* **33** 17
[5] Minami T, Oohashi K, Takata S, Mouri T and Ogawa N 1990 *Thin Solid Films.* **193** 721
[6] Nanto H, Minami T, Shooji S and Takata S 1984 *J. Appl. Phys.* **84** 1029
[7] Minami T 2005 *Semicond. Sci. Technol.* **20** 35
[8] T Minami H Sato H N and Takata S 1986 *Jap. J. Appl. Phys.* **25** 776
[9] A K Das P M and Kukreja L M 2009 *J. Phys. D: Appl. Phys* **42** 165405
[10] Lai L, Liu C, Lee C, Lou L, Yeh W and Chu M 2008 *J. Mater. Res.* **23** 2506
[11] Schofield M D 2011 master Thesis: *ZnO Deposited by Magnetron Sputtering Incorporating Si Nanoclusters*, University of Oslo, http://www.duo.uio.no/publ/fysikk/2011/ 127295/Masters_Thesis_-Matthew_Schofield.pdf
[12] Powell C J and Jablonski A 2000 NIST Electron Inelastic-Mean-Free-Path Database Version 1.1, National Institute of Standards and Technology, Gaithersburg, MD
[13] Cebulla R, Wendt R and Ellmer K 1998 *J. Appl. Phys.* **83** 1087
[14] Morkoç H and Ü Özgür 2009 *Zinc Oxide: Fundamentals, Materials and Device Technology* (Weinheim: Wiley-VCH)
[15] Birkholz M, Fewster P and Genzel C 2006 *Thin films analysis by X-ray scattering* (Weinheim: Wiley-VCH)
[16] Minami T, Nanto H, Sato H and Takata S 1988 *Thin Solid Films.* **164** 275
[17] Sato H, Minami T, Takata S, Mouri T and Ogawa N 1992 *Thin Solid Films.* **220** 327
[18] Deng Z, Huang C, Huang J, Wang M, He H, Wang H and Cao Y 2010 *J. Mater. Sci.: Mater. Electron.* **21** 1030
[19] Lu J G, Ye Z Z, Zeng Y J, Zhu L P, Wang L, Yuan J, Zhao B H and Liang Q L 2006 *J. Appl. Phys.* **100** 073714
[20] Agashe C, Kluth O, Hüpkes J, Zastrow U, Rech B and Wuttig M 2004 *J. Appl. Phys.* **95** 1911

Acceptor Doping in Sputtered ZnO Thin Films

K S Shtereva[1,2], V Tvarozek[1], I Novotny[1], P Sutta[3], A Vincze[4], S Flickyngerova[1] and M Vojs[1]

[1]Slovak University of Technology, Institute of Electronics and Photonics, Ilkovicova 3, 812 19 Bratislava, Slovakia
[2]University of Rousse, Department of Electronics, Studentska 8, 7017 Ruse, Bulgaria
[3]University of West Bohemia, New technologies-Research Center, 306 14 Plzen, Czech Republik
[4]International Laser Centre, Bratislava, Ilkovicova 3, 812 19 Bratislava, Slovakia

E-mail: KShtereva@ecs.uni-ruse.bg

Abstract. Acceptor doping of zinc oxide thin films prepared by radio-frequency diode sputtering, using nitrogen as a doping source, was realized during sputtering in Ar/N_2 gas mixture or via ion implantation of 180 keV N ions. Effects of N doping and (Al,Ga):N co-doping on the films' properties have been studied. Highly resistive ZnO:N films with *p*-type carrier concentrations ranging $10^{14}\div10^{15}$ cm^{-3} and more random orientation of the crystallites were obtained. The co-doped ZnO:Al:N and ZnO:Ga:N films maintained a *c*-axis texture in the direction of the surface normal. The carrier concentration of the *p*-type films was on the order of $10^{17}\div10^{18}$ cm^{-3}. 180 keV N ions were implanted in sputtered ZnO:Ga thin films. The implant doses varied $10^{15}\div2\text{x}10^{16}$ cm^{-2}. Annealing studies were performed under O_2 and N_2 ambient at different temperatures and times. N-implanted films were polycrystalline with a preferred *c*-axis orientation of the crystallites. Post-implantation annealing caused the increase of the (002) diffraction line integrated area and reduction in resistivity and the biaxial stress. After annealing in N_2, a *p*-type conduction was observed with a carrier concentration and average transmittance on the order of $2.8\text{x}10^{19}$ cm^{-3} and 81 % respectively.

1. Introduction

The demand for highly efficient light emitting diodes (LEDs) turned the attention to zinc oxide (ZnO) as an alternative LED material to GaN, due to its wide direct band gap of 3.37 eV and large exciton binding energy of 60 meV [1,2]. Decidedly, the application of ZnO for optoelectronic devices in the blue wavelength region depends on the creation of a good quality *n*- and *p*-type ZnO material. Despite many superior properties, the impossibility for controllable and reproducible *p*-type doping in ZnO has remained the major obstacle for the fabrication of ZnO-based devices. Among possible acceptor dopants, nitrogen has been recognized as a promising candidate for *p*-type doping in ZnO and there are reports on the realization of N-doped *p*-type ZnO [3-5]. Our earlier studies were focused on the use of N_2 as a doping source for the preparation of nitrogen-doped (ZnO:N) and (Al, Ga):N co-doped (ZnO:Al:N), (ZnO:Ga:N) thin films by radio-frequency (rf) diode sputtering in Ar/N_2 gas mixture at varying nitrogen contents [6-8]. The purpose of these studies was to investigate systematically effects of N incorporation and (Al,Ga):N co-incorporation on the electrical, structural and optical properties of sputtered ZnO thin films, by employing a number of characterization techniques, such as X-ray

Published under licence by IOP Publishing Ltd

diffraction measurements (XRD), Secondary Ion Mass Spectrometry (SIMS) and Atomic Force Microscopy (AFM), combined with Hall-effect and optical measurements. We found that nitrogen doping caused more random orientation of the crystallites in ZnO:N films deposited at 50 and 75 % N_2 in Ar/N_2 sputtering gas. XRD confirmed that co-doping improved the crystalline structure, and all ZnO:Al:N and ZnO:Ga:N thin films grew with a strong *c*-axis preferred orientation in direction of the surface normal. Nitrogen doping results in an increased absorption around the band edge, yellow to brown coloration and narrowing of the band gap ($E_g < 3.2$ eV) of the ZnO:N, ZnO:Al:N and ZnO:Ga:N films.

On the basis of these results and reports on *p*-type doping in ZnO thin films via nitrogen implantation [9,10], our current research has been focused on the acceptor doping via nitrogen implantation with the aim to overcome the solubility limitation and to introduce active N_O acceptors that can lead to *p*-type material.

In this work we summarize and compare some results and findings for N doping and (Al, Ga):N co-doping of ZnO films during sputtering in Ar/N_2 gas with those for N implanted ZnO:Ga thin films, annealed under O_2 and N_2 ambient. Our results suggest that doping via ion implantation can be more promising approach for *p*-type doping in ZnO. However, more studies are needed to understand phenomena and optimize parameters of the ion implantation processing (doping levels and doping uniformity) and post-implant annealing conditions.

2. Experimental details

ZnO:N, ZnO:Al:N and ZnO:Ga:N thin films were deposited on Corning 7059 glass substrates by rf diode sputtering of ceramic ZnO, ZnO:Al_2O_3 (98wt%:2wt%) and ZnO:Ga_2O_3 (98wt%:2wt%) targets using an argon/nitrogen (Ar/N_2) gas mixture. The percentage of nitrogen in the sputtering Ar/N_2 gas varied from 0 % to 100 %. The base pressure ($2x10^{-5}$ Pa), working gas pressure (1.3 Pa), and the sputtering power, 500 W (ZnO:N)/400 W (ZnO:Al:N)/600 W (ZnO:Ga:N), were maintained constant during the deposition process. The ZnO:N and ZnO:Al:N films were deposited on unheated substrates. ZnO:Ga:N thin films were prepared at a varying substrate temperature (T_S) that changed from room temperature (RT) to 300°C. The detailed description of the parameters of the sputtering process is given in Ref. [8,11,12]. Electrical properties were measured using a Hall system and software at a room temperature with a magnetic field of 0.76 T.

Recently, ion implantation was used to modify the properties of ZnO:Ga thin films deposited on Corning 7059 glass substrates by rf diode sputtering. The film thickness of ~ 500 nm was determined by Dektak 150 instrument. The ZnO:Ga films were *n*-type conductivity with an electron concentration of order ~ $7x10^{19}$ cm^{-3} and a mobility of ~ 1 cm^2/Vs measured from a Hall-effect system with a magnetic field of 0.15 T at a room temperature. Nitrogen ions were implanted into these films at 180 keV with doses of $1x10^{15}$, $5x10^{15}$, $1x10^{16}$ and $2x10^{16}$ cm^{-2}, under normal incidence. Monte Carlo simulations using the SRIM code were performed prior to implantation to obtain the theoretical nitrogen profiles for the various doses. The projected range (R_P) at 180 keV was determined to be 340 nm. Nitrogen-implanted ZnO:Ga thin films were annealed in oxygen (O_2) and in nitrogen (N_2).

The structure and film orientation of the ZnO:N thin films were evaluated by X-ray diffraction on AXS Bruker D8 powder diffractometer equipped with a 2D detector and an Eulerian cradle. CoKα radiation ($\lambda = 0.179$ nm) was used. For crystallinity characterization of the co-doped ZnO:Al:N, ZnO:Ga:N and N-implanted ZnO:Ga films, X-ray measurements were performed on X'pert Pro powder diffractometer (symmetric ϑ-ϑ geometry), equipped with an ultra-fast linear semiconductor detector PIXcel using CuKα radiation ($\lambda = 0.154$ nm). The method proposed by Langford [13], based on the XRD line profile analysis was used to perform the size-strain analysis (micro-strains and crystallite sizes). SIMS depth profiles were acquired with a TOF-SIMS IV analyzer (ION TOF GmbH, Muenster), using a Cs^+ primary ion beam with energy of 2 keV. Optical spectrophotometry measurements were carried out from the UV region to the near IR region by an Ava Spec-2048 Fiber Optic Spectrometer.

3. Results and discussions

Our experiments showed the profound effects of nitrogen doping on type and range of conductivity, the film microstructure and preferred orientation, the film composition and its optical properties, as well as the relationship structure-electrical & optical properties.

3.1. Analysis of doped and co-doped sputtered ZnO thin films

Highly resistive *p*-type ZnO:N thin films with the highest hole concentration of $2.6x10^{15}$ cm^{-3}, mobility of 2 cm^2/Vs and resistivity of $1.3x10^3$ Ωcm were obtained. XRD measurements revealed that N-doping and (Al,Ga):N co-doping in ZnO had a strong impact on the crystallographic texture of the deposited films. Three diffraction lines typical for ZnO (100), (002), and (101) appear in the XRD patterns of ZnO:N films (Figure 1 inset).

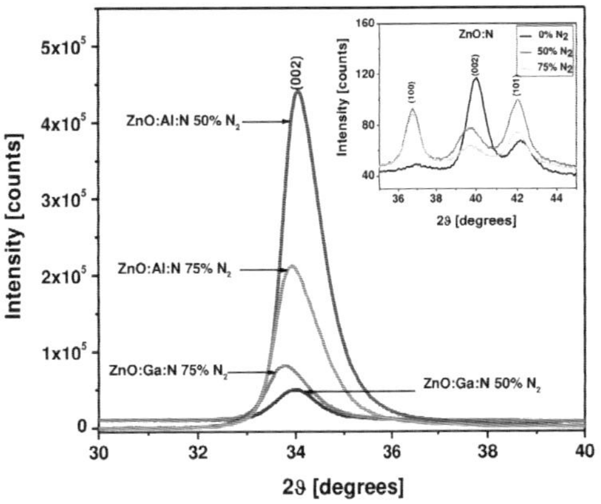

Figure 1. XRD patterns for ZnO:Al:N and ZnO:Ga:N films deposited with 50 and 75% N_2 in Ar/N_2 gas mixture. The inset shows XRD paterns for ZnO:N films.

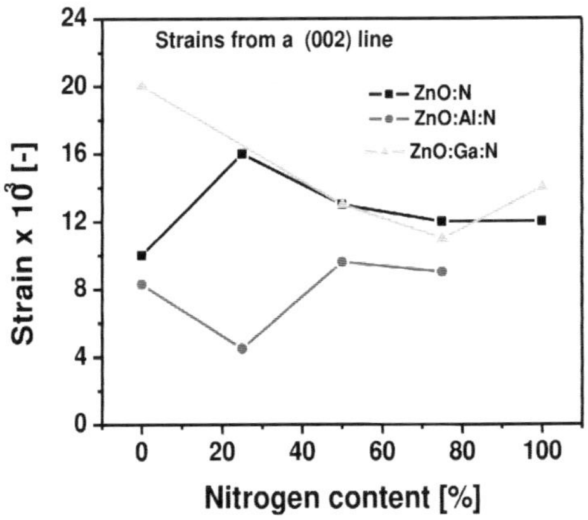

Figure 2. Micro-strains from a (002) diffraction line as a function of the N_2 content in Ar/N_2 gas mixture.

The diffraction lines' position, full width at half maximum (FWHM), and relative intensity varied with the N_2 content. A strong (002) line, and weak (100) and (101) features for un-doped ZnO:N films (0 % N_2, inset, black line), provide evidence for their polycrystalline hexagonal structure with a preferred c-axis orientation, perpendicular to the substrate. With the increasing N_2 content, the (002) line intensity decreased while the integrated intensities of the (100) and (101) lines increased. The dominant (100) diffraction line showed that the ZnO:N thin films deposited at 50% N_2 (inset, red line), and 75% N_2 (inset, green line), became more randomly orientated. Comparison of XRD reference data for hexagonal ZnO calculated by using APX 63 - Struc software [14] with those for ZnO:N show that the 2ϑ diffraction angles of N - doped films are slightly smaller than the standard values (37.09°, 40.22°, 42.39°), for (100), (002) and (101) lines of un-doped ZnO respectively. According the Bragg law the decrease of the diffraction angle with increasing N_2 content will increase the interplanar spacing *d* thus introducing stress into the film. Therefore, this shift can be explained by compressive lattice strains (stresses) created during sputtering process.

In contrast to them, all co-doped ZnO:Al:N and ZnO:Ga:N films exhibited a *c*-axis preferred orientation of the crystallites (Figure 1). The shift and the asymmetry of the (002) diffraction line are mostly due to interstitial oxygen and nitrogen atoms in the zinc oxide lattice structure. Because oxygen and nitrogen have substantially less atomic radius than zinc, in case of their redundancy in the lattice structure, they can create interstitial solid solutions. As a consequence of that, the lattice parameters of the ZnO structure increase and compressive lattice stresses are observed (peak position of the line is moving towards lower 2ϑ angle).

On the other hand, in case of ZnO:Al:N films, non-stoichiometric Al_xO_y and Al_xN_y minority phases can also appear, when the films are deposited by sputtering. They have very strong diffraction lines near to (002) ZnO line (on the right hand side of the (002) ZnO line), which leads to the asymmetry of the (002) line profile. The sharp Al_xO_y and Al_xN_y lines cannot be seen because there is only a small amount of that minority phases. Furthermore, their coherently diffracting domains are too small (in a nanometer scale), which leads to the broadening of the lines. They overlap each other so that it creates a smooth asymmetric line profile.

A ZnO powder standard was used for identifying the diffraction lines peak position. According to it, the peaks at 2ϑ of 31.77°, 34.42° and 36.25° are attributed to (100), (002) and (101) crystal planes of ZnO. Comparing this to the obtained 2ϑ values for the ZnO:Al:N films (34.06°/50%N_2, 33.93°/75%N_2) and the ZnO:Ga:N films (34°/50%N_2, 33.74°/75%N_2), indicates a shift towards lower angles of 2ϑ and compressive lattice strains in these films. In figure 2 are compared micro-strains from a (002) line for doped and co-doped thin films as a function of the N_2 content. The improved crystalline structure of the co-doped films resulted in better electrical properties. The ZnO:Al:N deposited with 75% N_2 showed *p*-type conduction with a hole concentration of 7.8 x 10^{17} cm^{-3}, more than two orders of magnitude higher than that in the ZnO:N films, and resistivity of 21 Ωcm. In *p*-type ZnO:Ga:N films grown in pure N_2, resistivity fell to 1.5 Ωcm and the hole concentration increased to 2.6x10^{18} cm^{-3} [8].

Figure 3 compares optical transmittance spectra of un-doped ZnO films (0% N_2) and ZnO:N, ZnO:Al:N and ZnO:Ga:N films deposited with 75% N_2 in the sputtering gas.

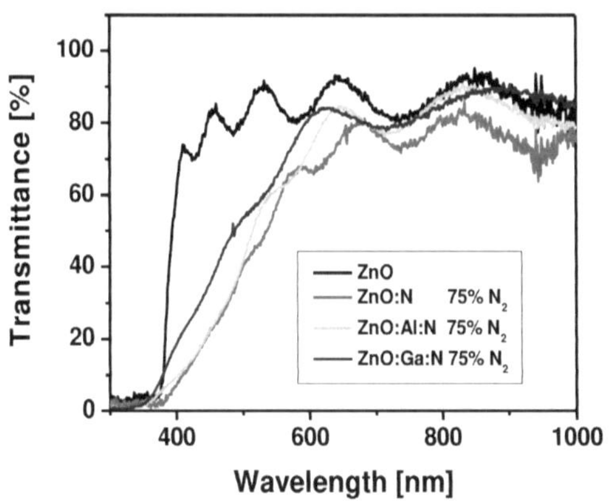

Figure 3. Optical transmittance spectra of un-doped ZnO films (0% N_2) and ZnO:N, ZnO:Al:N and ZnO:Ga:N samples deposited with 75% nitrogen in Ar/N_2 gas mixture.

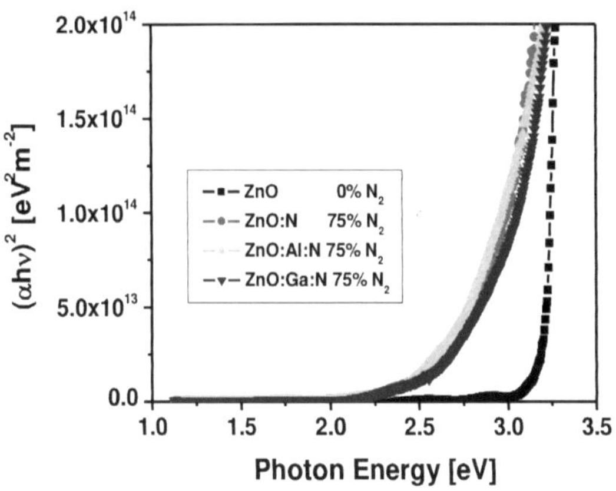

Figure 4. Plot of the absorption squared $(\alpha h\nu)^2$ versus phonon energy of un-doped ZnO films (0% N_2) and ZnO:N, ZnO:Al:N and ZnO:Ga:N films deposited with 75% N_2 in Ar/N_2 gas mixture.

All films showed light absorption into the visible region and had yellow colour due to N incorporation in the films. An increased absorption around the band edge and the yellow coloration associated with nitrogen doping was also observed by other research groups [15, 16] and has been assigned to deep N-related states in the band gap of ZnO. The transmittance data were used to calculate the absorption coefficients and to determine the optical band gap. ZnO is known as a direct band gap material [17] therefore, the direct electron transition from valence to conduction bands was assumed for these calculations (Figure 4). The optical band gap of ZnO (0% N_2) is $E_g \sim 3.24$ eV and narrows with doping and co-doping. The evaluated optical band gap of ZnO:Al:N films is 2.62 eV/50%N_2 and 2.73 eV/75%N_2. For ZnO:Ga:N films E_g was evaluated to be 2.89 eV and 2.94 eV for films deposited with 50 and75%N_2 respectively.

3.2. Analysis of N-implanted ZnO:Ga thin films

A dominant (002) diffraction line in the XRD patterns of N-implanted ZnO:Ga films, shows a *c*-axis preferred orientation of the crystallites in these films (Figure 5). Post-implantation annealing reduces the structure damage. In figure 6 is plotted the ratio of integrated intensities of the (002) and (101) lines, ($I_{(002)}/I_{(101)}$). The higher $I_{(002)}/I_{(101)}$ ratio suggests more textured structure after annealing in O_2 then after N_2 anneal. Similar trends were reported previously for N-implanted ZnO films [10,18].

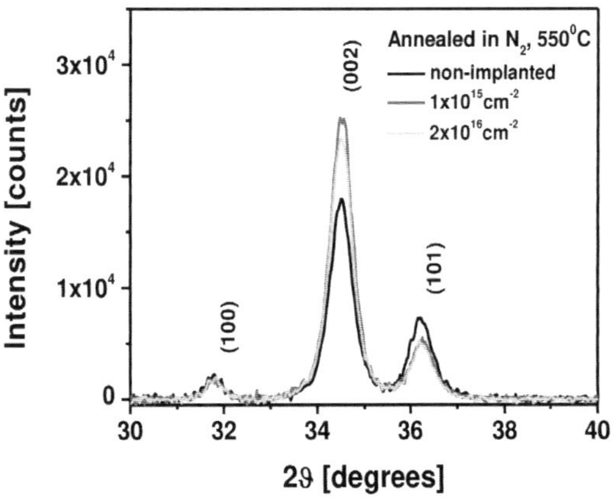

Figure 5. X-ray diffraction patterns for ZnO:Ga films, non-implanted and implanted with different doses, annealed in N_2 at 550°C for 10s.

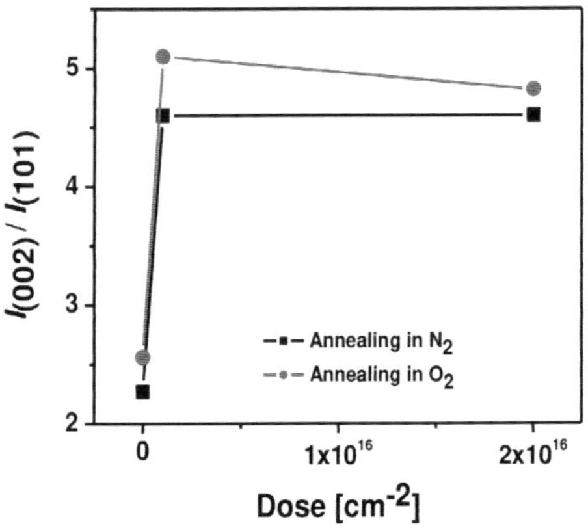

Figure 6. The ratio $I_{(002)}/I_{(101)}$ of integrated intensities of the (002) and (101) lines as a function of the implant dose.

We performed SIMS measurements on N-implanted ZnO:Ga films, before and after annealing, in order to study implantation and diffusion behavior of nitrogen in these films (Figure 7).

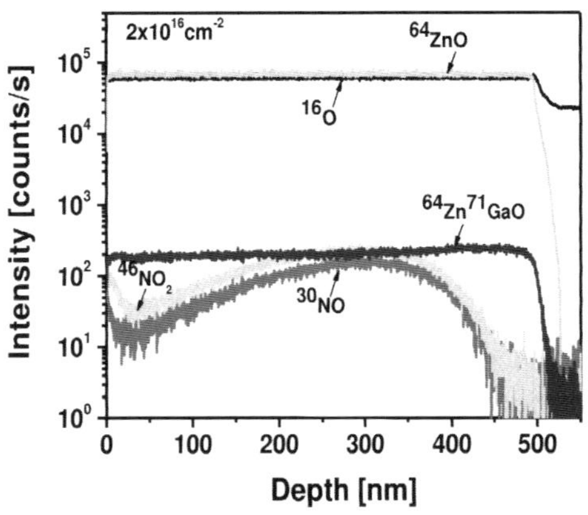

Figure 7. SIMS depth profiles of not annealed ZnO:Ga films implanted with $2x10^{16}$ cm^{-2}.

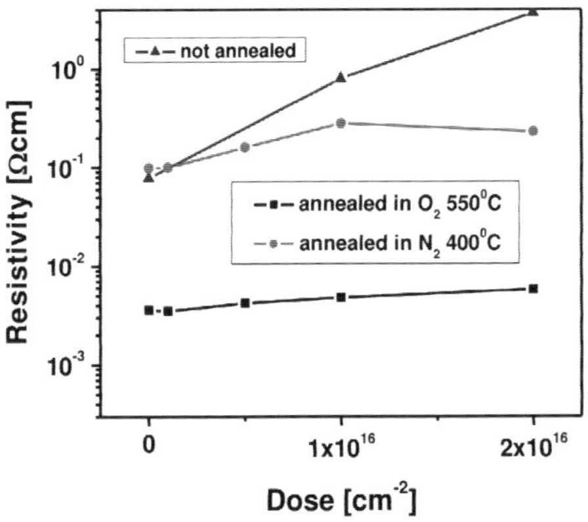

Figure 8. Dependence of resistivity on implant dose and annealing conditions.

The complex $^{30}NO^-$ and $^{46}NO_2^-$ ions are used to estimate nitrogen incorporation, since the negative $^{14}N^-$ ion produces a constant noise level intensity due to low ionization [19]. The distribution of complex $^{30}NO^-$ and $^{46}NO_2^-$ ions before annealing and after annealing in O_2 is Gaussian, and is in a good agreement with the simulated nitrogen profile from SRIM.

The non-implanted ZnO:Ga samples were *n*-type as determined by Hall-effect measurements and have almost uniform parameters, resistivity of $7.6x10^{-2}$ Ωcm and a carrier concentration of $7x10^{19}$ cm^{-3}. The resistivity of the films increased after nitrogen implantation providing evidence for the formation of N_O acceptors that partially compensate background donors in the ZnO:Ga films (Figure 8). The films' resistivity fell after annealing under both O_2 and N_2 ambient mainly as a result of the increased mobility (Table 1). In reference [20] N-implanted and annealed ZnO:Ga thin films were utilized to fabricate a metal–insulator–semiconductor ZnO-based diode.

TABLE 1. Electrical parameters of ZnO:Ga/ZnO:Ga:N films, as a function of the annealing conditions

Dose [cm^{-2}]	**Annealing conditions**	**Carrier concentration** [cm^{-3}]	**Carrier mobility** [cm^2/Vs]	**Resistivity** [Ωcm]	**Carrier type**
non-implanted	not annealed	$7x10^{19}$	1.2	$7.6x10^{-2}$	n
non-implanted	annealed in O_2 at 550°C	$7.5x10^{19}$	23	$3,6x10^{-3}$	n
$2x10^{16}$	not annealed	$4x10^{17}$	4.2	3.7	n
$2x10^{16}$	annealed in O_2 at 550°C	$5.1x10^{19}$	21	$5.8x10^{-3}$	n
$1x10^{16}$	annealed in O_2 at 550°C	$6x10^{19}$	21	$4.8x10^{-3}$	n
$2x10^{16}$	annealed in N_2 at 400°C	$9x10^{19}$	0.3	0.23	n
$1x10^{16}$	annealed in N_2 at 400°C	$2.8x10^{19}$	0.6	0.46	p

In contrast to previous work [9], a conversion of the conduction type was not observed after annealing in O_2, and all samples remained *n*-type as reported for N-implanted ZnO films in reference [21]. Higher Hall mobility in these films resulted from the better crystalline structure, the reduction of the defects and the stress in the films. A *p*-type carrier conduction (a hole concentration of $2.8x10^{19}$ cm^{-3}) was measured in a sample implanted with a $1x10^{16}$ cm^{-2} dose after annealing in N_2.

In figure 9 are compared optical transmittance spectra (including glass substrate) of the N-implanted ZnO:Ga films annealed in a N_2 atmosphere (400°C, 15 s) and in an O_2 atmosphere (550°C, 30 min).

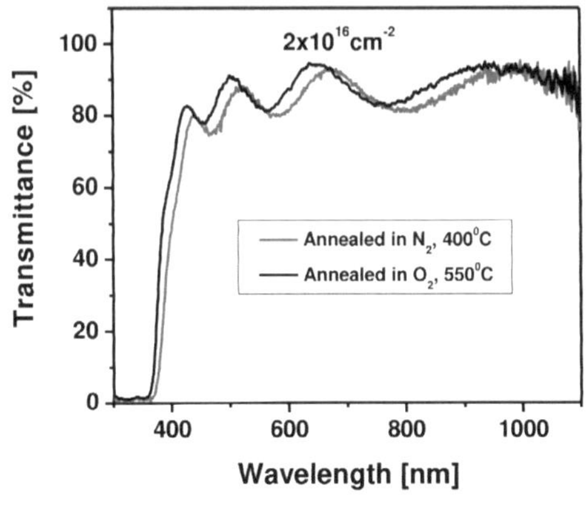

Figure 9. Optical transmittance spectra of N-implanted ZnO:Ga films annealed in N_2 and O_2.

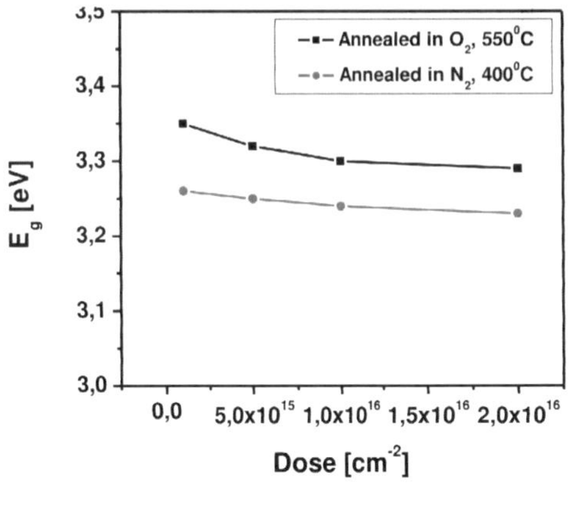

Figure 10. Optical band gap of N-implanted ZnO:Ga films as a function of the implant dose and the annealing conditions.

The transmittance data were acquired over the wavelength range of $200 < \lambda < 1100$ nm. All N-implanted films are highly transparent with an average transmittance > 84 % in the wavelength range of 390-1100 nm. It was found that optical transmittance depended on both, the annealing conditions and the implantation dose. Average transmittance of 87 % obtained after annealing in O_2 at 550°C resulted from better film crystallinity. For the samples annealed in N_2, average transmittance decreased with the increasing N-dose. Figure 10 shows the variations in the optical band gap with the implant dose and depending on the annealing conditions.

4. Conclusions

We attempted acceptor doping in ZnO via: (1) N-doping and (Al, Ga):N co-doping of sputtered ZnO thin films and (2) N implantation in sputtered ZnO:Ga thin films. It was found that microstructure, electrical and optical parameters strongly depend on N-doping.

XRD patterns of ZnO:N revealed fine grain textured films. The preferred orientation of the crystallites changed with an increasing N_2 content. Highly resistive *p*-type ZnO:N films with the hole concentrations $10^{14}\div10^{15}$ cm^{-3} and an average transmittance of 54% were obtained. Co-doping improved the crystalline structure. All ZnO:N:Al and ZnO:Ga:N thin films showed a *c*-axis texture. The shift and asymmetry of the (002) diffraction line are mostly due to interstitial oxygen and nitrogen atoms in the zinc oxide lattice structure. In case of ZnO:Al:N films also non-stoichiometric Al_xO_y and Al_xN_y minority phases can appear which leads to the asymmetry of the (002) line profile.The highest *p*-type carrier concentration of 2.6×10^{18} cm^{-3} corresponded to the ZnO:Ga:N films with a strong (002) preferred orientation and a grain size columnar structure, deposited in pure N_2.

Nitrogen ions were implanted in sputtered ZnO:Ga thin films at 180 keV with the doses ranging from 1×10^{15} to 2×10^{16} cm^{-2}. In N-implanted ZnO:Ga the preferred orientation of the crystallites was in (002) plane. Post-implant annealing increased the (002) texture and reduced the biaxial stresses in the film. The SIMS depth profile of the complex $^{30}NO^-$ ions is Gaussian. It was found that the annealing conditions and the temperature gradient had a profound impact on the film properties. The resistivity of the films increased after nitrogen implantation and it decreased significantly after post-implantation annealing in O_2 mainly as a result of the increased mobility and the improved crystallinity of the films. Depending on the annealing conditions, *p*-type material was obtained with a hole concentration of 2.8×10^{19} cm^{-3}, mobility of 0.6 cm^2/Vs, resistivity of 0.46 Ωcm in the sample implanted with a 1×10^{16} cm^{-2} dose after annealing in N_2 at 400°C. The N implanted ZnO:Ga thin films were highly transparent with an average transmittance higher than 84 % in the wavelength range of 390-1100 nm. Therefore, such films are more suitable for transparent electronics devices applications.

Acknowledgements

Presented work was supported by the the CENTEM project, reg. no. CZ.1.05/2.1.00/03.0088 cofunded from the ERDF the OP RDI programe of the MEYS CR within the project No. 1M06031 and by the SK projects VEGA 1/0459/12, 1/0787/09 and SRDA No. LPP-0094-09. Realization of ion implantation at the Ion Beam Centre, the University of Surrey, England, is also acknowledged.

References

[1] Look D C, Coşkun C, Claflin B and Farlow G C 2003 *Physica B* **340–342** 32
[2] Tüzemen, S and Gür, E 2007 *Optical Materials* **30** 292
[3] Li X, Asher S E, Limpijumnong S, Keyes B M, Perkins C L, Barnes T M, Moutinho H R, Luther J M, Zhang S B, Wei S-H and Coutts T J 2006 *J. Cryst. Growth* **287** 94
[4] Zaidi T, W E Fenwick, Melton A, Li N, Gupta Sh, Yu H, Ougazzaden A and Ferguson I 2008 *J. Cryst. Growth* **310** 5011
[5] Tvarozek V, Shtereva K, Novotny I, Kovac J, Sutta P, Srnanek R and Vincze A 2008 *Vacuum* **82** 166
[6] Tvarozek V, Novotny I, Sutta P, Flickyngerova S, Schtereva K and Vavrinsky E 2007 *Thin Solid Films* **515** 8756
[7] Vincze A, Kováč J, Novotný I, Bruncko J, Haško D, Šatka A and Shtereva K 2008 *Appl. Surf. Sci.* **255** 1419

[8] Shtereva K, Flickyngerova S, Tvarozek V, Novotny I, Kovac J and Vincze A 2011 *Vacuum* doi:10.1016/j.vacuum.2011.07.006
[9] Georgobiani N, Gruzintsev A N, Volkov V T, Vorobiev M O, Demin V I and Dravin V A 2003 *Nuclear Instr. & Methods in Phys. Research* **A 514** 117
[10] Lin Ch-Ch, Chen S-Y, Cheng S-Y and Lee H-Y 2004 *Appl. Phys. Lett.* **84** (24) 5040
[11] Shtereva K, Novotny I, Tvarozek V, Sutta P, Vincze A and Pullmannova A 2010 *J. Electrochem. Soc.* **157 (9)**, H891
[12] Shtereva K S, Sutta P, Pecz B, Novotny I, Kovac J and Tvarozek V 2011 *ECS Transactions* **35 (6)** 141
[13] Langford J I 1978 *J. Appl. Crystallography* **11** 10
[14] Kraus I 1993 in *Struktura a vlastnosti krystalů* (Academia, Praha) p. 221
[15] Kobayashi K, Kondo Y, Tomita Y, Maeda Y and Matsushima S 2007 *Appl. Surf. Sci.* **253** 5035
[16] Wang J, Sallet V, Jomard F, Botelho do Rego A M, Elamurugu E, Martins R and Fortunato E 2007 *Thin Solid Films* **515** 8785
[17] Sze S M 1981 *Physics of Semiconductor Devices* 2nd edition (N. Y. John Wiley & Sons, Inc.) 72
[18] Yang Ch-Ch, Lin Ch-Ch, Peng Ch-H and Chen S-Y 2005 *J. Cryst. Growth* **285** 96
[19] Komatsu M, Ohashi N, Sakaguchi I, Hishita S and Haneda H 2002 *Appl. Surf. Sci.* **189** 349
[20] Alivov Ya I, Look D C, Ataev B M, Chukichev M V, Mamedov V V, Zinenko V I, Agafonov Yu A and Pustovit A N 2004 *Solid-State Electronics* **48** 2343
[21] Vaithianathan V, Hishita Sh, Moon J H and Kim S S 2007 *Thin Solid Films* **515** 6927

The effect of Substrate temperature on physical and electrical properties of DC magnetron sputtered $(Ta_2O_5)_{0.85}(TiO_2)_{0.15}$ films

M. Chandra Sekhar[1], S. Uthanna[1], R. Martins[2], S.V. Jagadeesh Chandra[2] and E. Elangovan[2]

[1]Department of Physics, Sri Venkateswara University, Tirupati – 517 502, India

[2]CENIMAT/I3N, Departamento de Ciência dos Materiais, Faculdade de Ciênciase Tecnologia, FCT, Universidade Nova de Lisboa and CEMOP/UNINOVA, 2829-516 Caparica, Portugal

E-mail: uthanna@rediffmail.com

Thin films of $(Ta_2O_5)_{0.85}(TiO_2)_{0.15}$ were deposited on quartz and p-Si substrates by DC reactive magnetron sputtering at different substrate temperatures (T_s) in the range 303 – 873 K. The films deposited at 303 ^{0}K were in the amorphous and it transformed to crystalline at substrate temperatures ≥ 573 ^{0}K. The crystallite size was increased from 50 nm to 72 nm with the increase of substrate temperature. The surface morphology was significantly influenced with the substrate temperature. After deposition of the $(Ta_2O_5)_{0.85}(TiO_2)_{0.15}$ films on Si, aluminium (Al) electrode was deposited to fabricate metal/oxide/semiconductor (MOS) capacitors with a configuration of Al/$(Ta_2O_5)_{0.85}(TiO_2)_{0.15}$/Si. A low leakage current of 7.7×10^{-5} A/cm^2 was obtained from the films deposited at 303 K. The leakage current was decreased to 9.3×10^{-8} A/cm^2 with the increase of substrate temperature owing to structural changes. The conduction mechanism of the Al/$(Ta_2O_5)_{0.85}(TiO_2)_{0.15}$/Si capacitors was analyzed and compared with mechanisms of Poole-Frenkel and Schottky emissions. The optical band gap (E_g) was decreased from 4.45 eV to 4.38 eV with the increase in substrate temperature.

1. Introduction

The continuous miniaturization of microelectronic devices requires high quality dielectric materials with high permittivity. Several high-*k* metal oxides such as titanium oxide (TiO_2), zirconium oxide (ZrO_2), hafnium oxide (HfO_2), and tantalum oxide (Ta_2O_5) [1–4] were extensively investigated for the replacement of basic Si based dielectrics. Among those high-k materials, Ta_2O_5 has been widely used as a gate dielectric material because of its high dielectric constant and a high degree of compatibility with the components of microelectronic devices [5]. Ta_2O_5 can accommodate a wide variety of doped ions through subtle variation of its crystal structure. It has been reported that the electrical properties of Ta_2O_5 can be improved through 8% substitution of TiO_2 [6]. In this work DC magnetron sputtering technique was employed for preparation of $(Ta_2O_5)_{0.85}(TiO_2)_{0.15}$ thin films and examined their physical properties at different substrate temperatures. MOS capacitor with configuration of Al/$(Ta_2O_5)_{0.85}(TiO_2)_{0.15}$/p-Si was fabricated and the effect of substrate temperature on the leakage currents were systematically investigated. Sputtering is an attractive technique in thin film

Published under licence by IOP Publishing Ltd

technology because of several advantages such as good homogeneity, uniform deposition over large area and good adhesion to the surface of substrates [7].

2. Experimental

Thin films of $(Ta_2O_5)_{0.85}(TiO_2)_{0.15}$ were deposited on quartz and 15 Ω cm (100) p-Si (having a boron doping concentration of $0.2x10^{15}$ – $2.2x10^{15}$ atoms/cm^3) substrates by DC reactive magnetron sputtering technique. The p-Si substrates were rinsed in deionized water and then immersed in 5 % HF solution to remove any native oxide present on the surface. It was followed by rinsing in deionised water and then flushed with dry Nitrogen prior to loading in to the deposition chamber. Pure tantalum and titanium was used as a sputter mosaic target. The vacuum pumping system was combination of diffusion and rotary pumps. Pure oxygen and argon were used as reactive and sputter gases respectively. The films were deposited with an oxygen partial pressure (OPP) of 6×10^{-2} Pa at different substrate temperatures (T_s) in the range 303 – 873 K, and at sputtering pressure of 5×10^{-1} Pa. X-ray photoelectron spectroscopy (XPS) (Phoibos 100MCD Energy Analyzer) and X-ray diffraction (XRD) (Seifert model 3003 TT) studies were used to analyze the composition and crystallographic structure of the films. The thickness of the films deposited on silicon and quartz substrates (400 ± 5 nm) were measured by using α- step profilometer. The surface microstructure of the films was analyzed by using the atomic force microscopy (AFM) (Digital Instrument, 3100 series). The deposited films were used to fabricate MOS capacitors with the configuration of $Al/(Ta_2O_5)_{0.85}(TiO_2)_{0.15}$/p-Si. An aluminum layer deposited by vacuum evaporation (area of 0.3 mm × 0.3 mm) was used as the top electrode and ohmic contact at the back-side of the p-Si substrate. The current - voltage (I-V) characteristics of the capacitors were measured by using Hewlett Packard pA meter. The optical transmittance spectra of the films deposited on quartz substrates were recorded by using UV-Vis-NIR double beam spectrophotometer.

3. Results and discussion

Figure 1 shows the representative XPS survey scan obtained from $(Ta_2O_5)_{0.85}(TiO_2)_{0.15}$ thin films deposited at 303 K. The survey spectrum shows the characteristic tantalum, titanium, oxygen and carbon peaks. The presence of carbon peak is related to organic surface contamination, which corresponds to the fact that the samples had been exposed to air before measurements. Carbon peak was disappeared from the surface of TiO_2 films after argon sputtering. Figure 2 shows the XPS core level binding energy spectra of Ta 4f, Ti 2p and O 1s films. Figure 2a shows the XPS core level binding energy spectra of Ta 4f of Ta_2O_5 The core level binding energies of Ta 4f gives raise to two peaks at 26.2 and 28.0 eV that correspond to Ta $4f_{7/2}$ and Ta $4f_{5/2}$, respectively, with spin-orbit splitting of 1.8 eV indicated the presence of Ta in the oxidation state of Ta^{5+} [7]. The core level binding energy of Ta $4d_{5/2}$ and Ta $4d_{3/2}$ were 230.4 and 241.9 eV, respectively. Figure 2b shows the XPS core level binding energy spectra of Ti 2p of TiO_2 The core level binding energy of Ti $2p_{3/2}$ and Ti $2p_{1/2}$ were 458.7 and 464.3 eV, with spin-orbit splitting of 5.6 eV indicated the presence of titanium in the oxidation state of Ti^{4+} [8]. Figure 2c shows the XPS core level binding energy spectra of O 1s. The O 1s core level binding energy is observed at 530.2 eV. The Ta/ Ti ratio in the films estimated from the XPS peak area showed nearly stoichiometric $(Ta_2O_5)_{0.85}(TiO_2)_{0.15}$ films.

The XRD patterns obtained from the $(Ta_2O_5)_{0.85}(TiO_2)_{0.15}$ films deposited on p-Si substrates are comparatively shown in Fig. 3 as a function of T_s. The films deposited at 303 K were amorphous. When the T_s increased to 573 K, the diffractions from (0 0 1), (1 11 0), (2 0 1), (0 0 2) and (3 11 1) planes were obtained at 2θ of 22.9^0, 28.3^0, 37.1^0, 46.8^0 and 56.2^0, respectively, related to orthorhombic Ta_2O_5 [JCPDS card No. 25-0922]. It is noteworthy that no diffraction peak was evidenced from impurity dopant (TiO_2), which is presumably due to its low content. Apparently, the intensity of (001) orientation had been prominent over other orientations. The intensity of the diffracted peaks is increased with further increase in T_s to 873 K. The full width-at-half-maximum (FWHM) of the XRD peak (001) is decreased from 0.123° to 0.110° with increasing T_s, which implies that the lower T_s might reduce their crystallite sizes. The crystallite size of the films is increased from 50 nm to 72 nm with the increase in T_s from 573 K to 873 K. The increase in crystallite size with increase in T_s is attributed to the atomic, molecular or ionic species of Ta_2O_5 impinged on the substrate surface acquire large thermal energy and hence large mobility. As a result, a large number of nuclei formed which coalescence at higher temperature to form a uniform film with good adhesion.

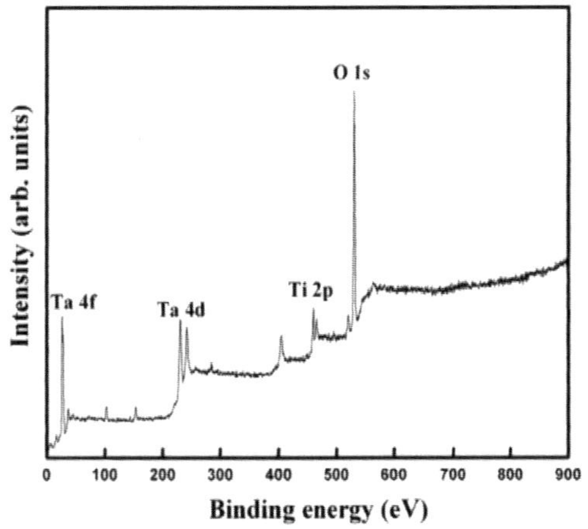

Fig.1 XPS survey scan of $(Ta_2O_5)_{0.85}(TiO_2)_{0.15}$ films

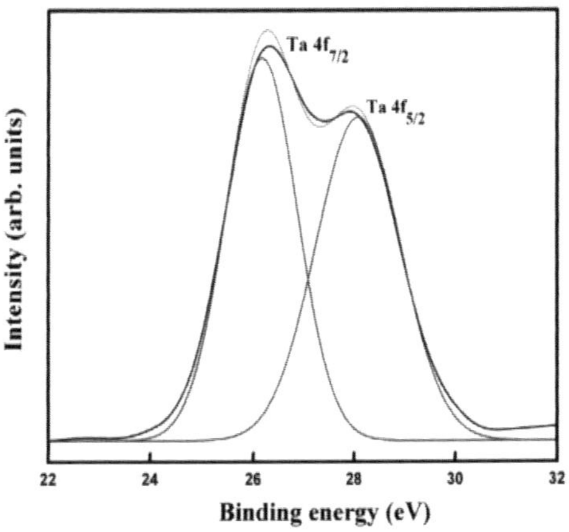

Fig. 2a Narrow-scan XPS spectra of Ta 4f

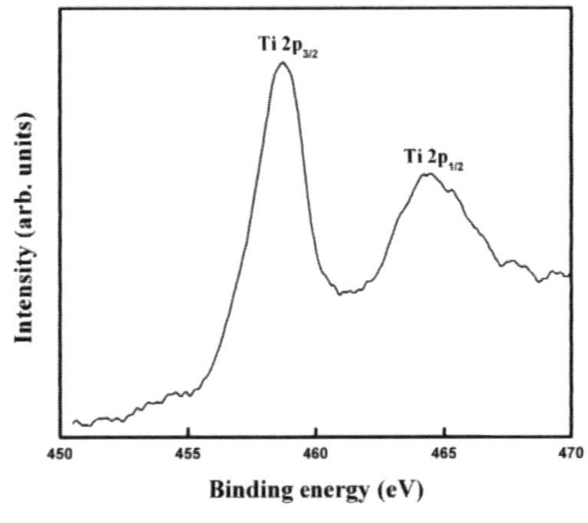

Fig. 2b Narrow-scan XPS spectra of Ti 2p

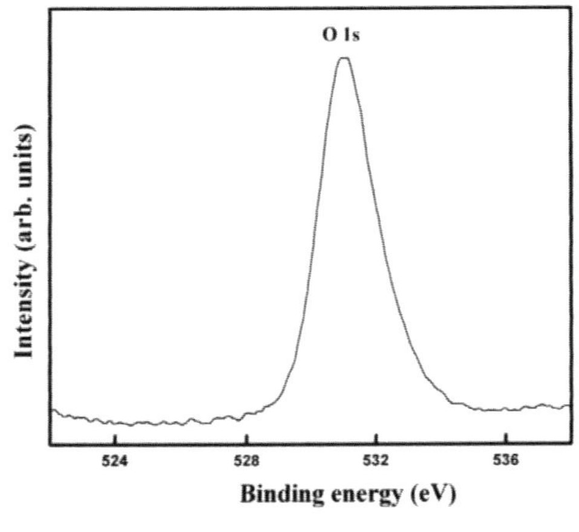

Fig. 2c Narrow-scan XPS spectra of O 1s

AFM microstructures obtained from the $(Ta_2O_5)_{0.85}(TiO_2)_{0.15}$ films deposited on silicon substrates at different T_s are shown in Fig. 4. The surface microstructure is significantly varied with the change in T_s. The microstructure obtained from the $(Ta_2O_5)_{0.85}(TiO_2)_{0.15}$ films deposited at 303 K showed featureless smooth surface, indicating amorphous nature, in accordance with the XRD studies. By increasing Ts, the surface of the films becomes rougher. Thus, increase of T_s causes a difference in the surface energy that improves the surface morphology of the growing films. The root mean square roughness (R_{rms}) is increased to 3.48 nm with the increase in T_s at 573 K there after it decreased to 1.78 nm.

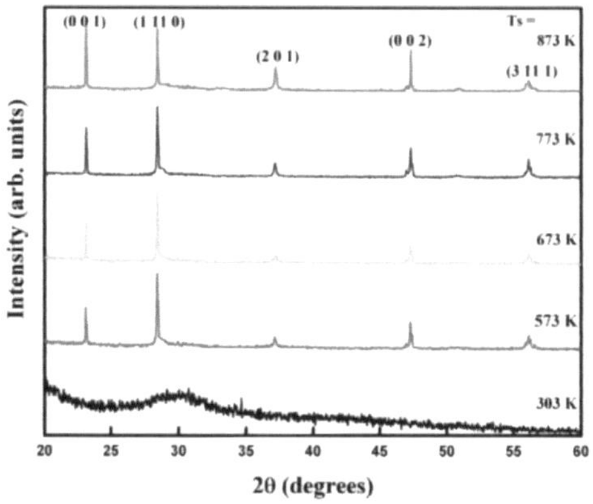

Fig. 3 XRD patterns of $(Ta_2O_5)_{0.85}(TiO_2)_{0.15}$ films.

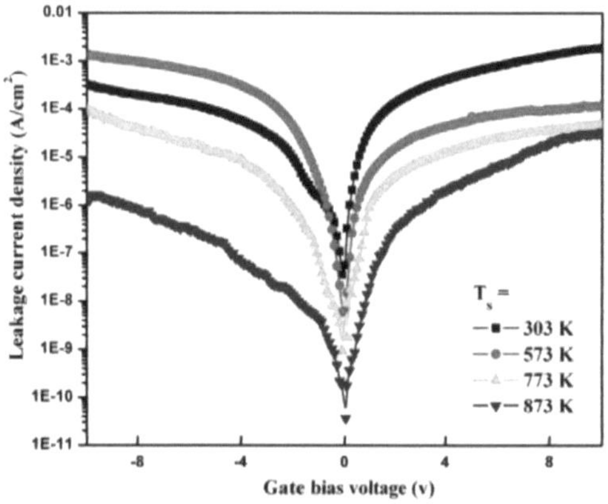

Fig. 5 J-V characteristics of the $Al/(Ta_2O_5)_{0.85}(TiO_2)_{0.15}/p$-Si capacitors.

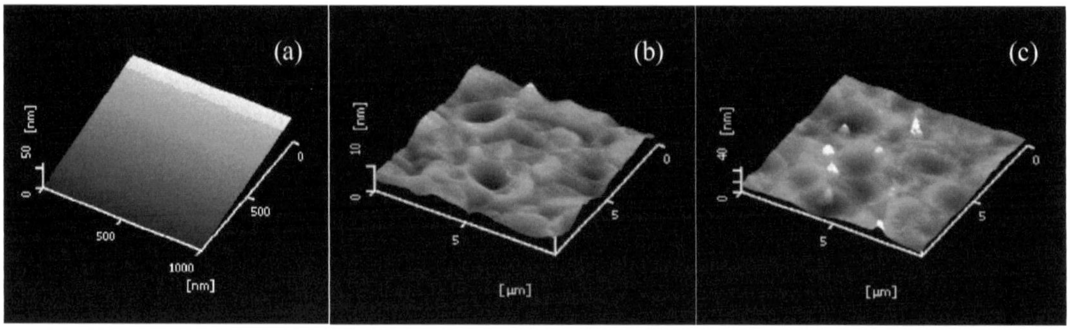

Fig. 4 AFM microstructures obtained from the films deposited with T_s : (a) 303 K, (b) 673 K and (c) 873 K

Figure 5 shows the J–V curves (for both +Ve and -Ve bias voltages) of the fabricated $Al/(Ta_2O_5)_{0.85}(TiO_2)_{0.15}/Si$ capacitors, where the $(Ta_2O_5)_{0.85}(TiO_2)_{0.15}$ thin films had been deposited at various T_s. The leakage current density obtained from the films deposited at 303 K was 7.7×10^{-5} A/cm^2 at a gate voltage of 1.5 V which decreased to a value of 9.3×10^{-8} A/cm^2 with the increase of T_s to 873 K [7]. The leakage current density decrease with the increase of T_s is attributed probably to the growth of denser and uniform grain boundary structure for the films deposited at elevated substrate temperatures. In addition, the improvement in crystallinity was also observed with increasing T_s. The leakage current is influenced by the grain size, crystallinity, and surface microstructure. When leakage current of $Al/(Ta_2O_5)_{0.85}(TiO_2)_{0.15}/Si$ capacitors is plotted as a function of electric field, it is observed that it follows Schottky emission at low electric fields and it may be Poole–Frenkel emission at higher electric fields. Figure 6 shows the plots of ln(J) versus $E^{1/2}$ for positive bias voltages. It is shown that the leakage currents are almost proportional to the electrical field [9]. Schottky emission (SE) occurs at the interface between the $(Ta_2O_5)_{0.85}(TiO_2)_{0.15}$ film and the Al electrode where a Schottky barrier is formed. Schottky emission is a field-assisted thermionic emission of an electron over a surface barrier. The SE conduction was given by the following equation [10]

$$J_{SE} = A^* T^2 \exp\left\{ -q\left(\frac{\Phi_B - \sqrt{qE/4\pi\varepsilon_0 k_r}}{kT} \right) \right\}$$

where $A^* = \frac{4\pi q m^* k^2}{h^3}$, the Richardson constant. J_{SE} the current density governed by Schottky emission, m^* the effective mass of electron. Φ_B, k_r, k, and h the Schottky barrier height, dynamic dielectric constant, Boltzmann's constant, and Planck's constant respectively.

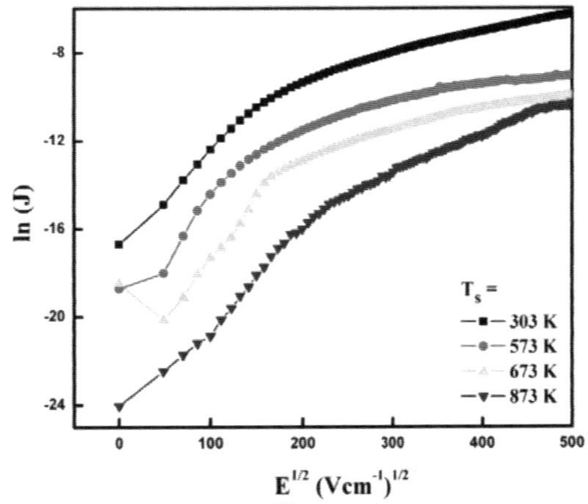

Fig. 6 Plots of $E^{1/2}$ versus ln(J) of $Al/(Ta_2O_5)_{0.85}(TiO_2)_{0.15}/p\text{-}Si$ capacitors

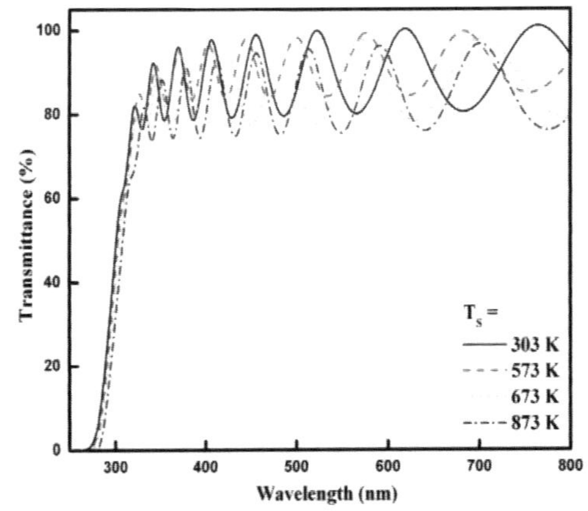

Fig. 7 Optical transmittance spectra of $(Ta_2O_5)_{0.85}(TiO_2)_{0.15}$ films

Figure 7 shows the optical transmittance spectra of $(Ta_2O_5)_{0.85}(TiO_2)_{0.15}$ films deposited at different T_s. It is seen that the optical transmittance of the films deposited at 303 K was the highest. When T_s increased to 873 K, the optical transmittance of the films slightly decreased. A sharp absorption edge was observed at the wavelength of about 300 nm, which is shifted towards higher wavelength with the increase in T_s. The optical band gap of the films deposited at 303 K was 4.45 eV, which is decreased to 4.38 eV with the increase in T_s to 873 K.

4. Conclusion

$(Ta_2O_5)_{0.85}(TiO_2)_{0.15}$ thin films were deposited on quartz and p-Si (100) substrates by DC magnetron sputtering at different substrate temperatures in the range of 303 K – 873 K. The X-ray photoelectron spectra showed that the atomic Ta/ Ti ratio is nearly stoichiometric in the $(Ta_2O_5)_{0.85}(TiO_2)_{0.15}$ films. The films deposited at 303 K were amorphous whereas those deposited at 573 K were transformed into crystalline and the crystallinity of the films was increased with further increase of substrate temperature up to 873 K. The surface morphology of the films was affected by the substrate temperature as confirmed by AFM experiments. The leakage current density of the $Al/(Ta_2O_5)_{0.85}(TiO_2)_{0.15}$ /p-Si structure was decreased from 7.7×10^{-5} A/cm^2 to 9.3×10^{-8} A/cm^2 with the increase in substrate temperature. There are two possible conduction mechanisms to explain the leakage currents in the $(Ta_2O_5)_{0.85}(TiO_2)_{0.15}$ thin films. Schottky effect was observed in the lower electric fields where as Poole-Frenkel effect was observed at higher electric fields. The optical band gap of the films deposited at 303 K was 4.45 eV, which is decreased to 4.38 eV with the increase of substrate temperature to 873 K.

Acknowledgement

This work was carried out with the financial support of University Grants Commission, New Delhi, India through a sanctioned Major Research Project: F. No. 34-36/2008 (SR) dated: 30-12- 2008.

One of the authors, M. Chandra Sekhar thankful to UGC for sanctioned UGC-RFSMS Fellowship. The authors would like to thank Prof. G. Mohan Rao, Department of Instrumentation and Applied Physics, Indian Institute of Science, Bangalore, India for his valuable suggestions and providing XPS data on the experimental thin films.

References

[1] M. Kadoshima, M. Hiratani, Y. Shimamoto, K. Torii, H. Miki, S. Kimura and T. Nabatame, *Thin Solid Films* **424**, 224, 2003.
[2] S. Ramanathan, C.M. Park and P.C. McIntyre, *J. Appl. Phys.* **91**, 4521, 2002.
[3] A. Callegari, E. Cartier, M. Gribelyuk, H.F.O. Schmidt and T. Zabel, *J. Appl. Phys.* **90,** 6466, 2001.
[4] G.B. Alers, D.J. Werder, Y. Chabal, H.C. Lu, E.P. Gusev, E. Garfunkel, T. Gustafsson and R.Urdahl, *Appl. Phys. Lett.* **73**, 1517, 1998.
[5] H. Shini and M. Nakata, *IEEE Trans. Electron Dev.* **38**, 455, 1991.
[6] J.Y. Gan, Y.C. Chang and T.B. Wu, *Appl. Phys. Lett.* **72,** 332, 1998.
[7] S.V. Jagadeesh Chandra, S. Uthanna and G. Mohan Rao, *Appl. Surf. Sci.* **254**, 1953, 2008.
[8] M. Chandra Sekhar, P. Kondaiah, S.V. Jagadeesh Chandra, G. Mohan Rao and S. Uthanna, *Appl. Surf. Sci.,* **258**, 1789, 2011.
[9] Y. Fukuda, K. Aoki and K. Numata, *Jpn. J. Appl. Phys.* **33**, 5255 1994
[10] S.M. Sze, *Physics of Semiconductor Devices*, 2nd ed. Wiley, New York, 402, 1981.

AUTHOR INDEX

Institute of Physics
Dirac House, Temple Back
Bristol BS1 6BE UK

proceedings.com